Algal
Cell
Motility

Current Phycology

Series Editors: Matthew J. Dring, Queen's University Belfast
Michael Melkonian, Universität zu Köln

Algal Cell Motility

Edited by
Michael Melkonian

Chapman and Hall
New York and London

First published in 1992 by
Chapman and Hall
an imprint of
Routledge, Chapman & Hall, Inc.
29 West 35 Street
New York, NY 10001-2291

Published in Great Britain by

Chapman and Hall
2-6 Boundary Row
London EC1 8HN

Library of Congress Cataloging in Publication Data

Algal cell motility / edited by Michael Melkonian.
     p.   cm.—(Current phycology)
   Includes bibliographical references and index.
   ISBN 0-412-02431-4
   1. Algae—Cytology.   2. Cells—Motility.   I. Melkonian, Michael,
  1948– . II. Series.
  QK565.A375   1991
  589.3'8764—dc20   91-17475
                                         CIP

British Library cataloging in publication data also available.

Contents

Contributors

Peter L. Beech—Universität zu Köln, Botanisches Institut, Lehrstuhl 1, Gyrhofstrasse 15, D-5000 Köln 41, Federal Republic of Germany

Stuart F. Goldstein—Department of Genetics & Cell Biology, 250 Biological Sciences Center, University of Minnesota, Minneapolis MN 55108, U.S.A.

Franz Grolig—Membran- und Bewegungsphysiologie, Botanisches Institut 1, Justus-Liebig Universität, Senckenbergstrasse 17, D-6300 Giessen, Federal Republic of Germany

Donat-P. Häder—Institut für Botanik und Pharmazeutische Biologie, Friedrich-Alexander-Universität, Staudtstr. 5, D-8520 Erlagen, Federal Republic of Germany

Egbert Hoiczyk—Institut für Botanik und Pharmazeutische Biologie, Friedrich-Alexander-Universität, Staudtstr. 5, D-8520 Erlangen, Federal Republic of Germany

Ritsu Kamiya—Department of Molecular Biology, School of Science, Nagoya University, Nagoya 464-01, Japan

Christos Katsaros—Institute of General Botany, University of Athens, Athens 15784, Greece

Michael Melkonian—Universität zu Köln, Botanisches Institut, Lehrstuhl 1, Gyrhofstrasse 15, D-5000 Köln 41, Federal Republic of Germany

Dorothee Schulze—Biologische Anstalt Helgoland, Wattenmeerstation List, Hafenstrasse 40, D-2282 List/Sylt, Federal Republic of Germany

Gottfried Wagner—Membran- und Bewegungsphysiologie, Botanisches Institut 1, Justus-Liebig Universität, Senckenbergstr. 17, D-6300 Giessen, Federal Republic of Germany

Richard E. Williamson—Plant Cell Biology Group, Research School of Biological Sciences, Australian National University, Canberra, A.C.T. 2601, Australia

Preface

Algae exhibit the greatest variety of cell motility phenomena in the living world. These range from the peculiar gliding motility of filamentous blue-green algae or cyanobacteria to chloroplast movements and cytoplasmic streaming which are most common in higher plants. In addition, cell motility by eukaryotic flagella is the characteristic mode of cell locomotion in algal flagellates and most reproductive cells of algae. Algae use these cell motility systems mainly to orient themselves or their photosynthetic organelles in a suitable light gradient to optimize growth and reproduction. In consequence most of the motility systems are coupled to photoreceptors and are regulated by signal transduction cascades. Algal cell motility has thus attracted considerable interest also from non-phycologists and some algal motility systems have become models of research in cell and molecular biology. This book summarizes some of the progress that has been made in recent years in the analysis of cell motility phenomena in the algae. Although complete coverage of the subject was not attempted, the six chapters cover all the major types of cell motility systems and the authors provide in depth reviews of gliding motility, chloroplast movements, cytoplasmic streaming, flagellar beat patterns, mechanisms of flagellar movement and centrin-mediated cell motility. The chapters reflect the different status of analysis of individual cell motility phenomena ranging from in depth cell and molecular analysis of model systems (chapters 3 and 5) to more comparative descriptions of a variety of cell motility phenomena (chapters 1 and 4). From the latter analyses it is obvious that many highly unusual cell motility phenomena exist in the algae that await a detailed analysis using the now available repertoire of cell and molecular techniques. If this book can stimulate interest toward this end, it has fulfilled its purpose.

I like to express my thanks to the contributors of the chapters who have made my job extremely easy. Also I like to thank Peter L. Beech (Melbourne) and Barbara Surek (Cologne) for spending many hours in reading chapters, making invaluable suggestions and helping in the preparation of the index.

Cologne, September 1991

Chapter 1

Gliding Motility

Donat-P. Häder and Egbert Hoiczyk

Introduction

Active motility is one of the fundamental characteristics of many microorganisms and involves the interplay of a number of cellular functions that enable an organism to move in a coordinated fashion. Therefore, cell motility is a very complex and broad topic in research, one that involves problems associated with motor design, steering and control, and energy supply and distribution. As early as the Precambrian era, organisms in aquatic habitats developed two basic types of motility: flagellar and gliding motility. The first one is the subject of other chapters in this volume; the second one is the topic of this chapter.

Microorganisms employ motility to optimize the position in their microhabitat using a number of external chemical and physical factors. Most motile gliding organisms use different environmental stimuli, such as photic (Nultsch, 1971; Nultsch and Wenderoth, 1973; Nultsch and Häder, 1979; Haupt, 1983), chemical (Fechner, 1915; Drews, 1959; Hopkins, 1969; Kaiser et al., 1979; Ho and McCurdy, 1979; Dworkin, 1983), galvanic (Verworn, 1889; Häder, 1977), mechanical (Schmid, 1918; Wagner, 1934; Williams, 1965), gravitational (Hopkins, 1966; Harper, 1976; Bean, 1984), and thermal (Reimers, 1928; Castenholz, 1968) clues to direct their search for a suitable niche for survival and growth in the biosphere.

Although gliding motility was observed more than two centuries ago (Adanson, 1767), the underlying basic mechanisms are still obscure in most organisms and pose a major biological research problem. Currently, the only generally accepted statement is that the mechanism of gliding movement differs from one taxonomic group to the next, so that the simple sounding question has to be answered for each organism individually: How does gliding motility work? Gliding seems to have been developed several times during evolution. Furthermore, modifications of the basic patterns resulted in a host of different mechanisms among organisms with distinct features. Therefore, gliding motility is only superficially a common characteristic of different taxonomic groups.

General Aspects

In contrast to swimming, gliding can be defined as an active translocation of an organism in contact with a solid or semisolid substrate or even through a highly viscous matrix (Halfen and Castenholz, 1971) without a microscopically detectable organelle for locomotion or (in most cases) a visible change in shape (Jarosch, 1962). Additional characteristics can be secretion of mucilage, frequent changes of direction (so-called to-and-fro movements) and the tendency to move in well-defined patterns like swarms or comets (Fig. 1.1). In recent reports the terms *sliding, twitching, jerking*, and *swarming* are also used to describe the motility phenomena (Henrichsen, 1972; Burchard, 1981), but they always emphasize distinct aspects rather than the general phenomenon.

Gliding is widespread throughout different phylogenetic taxa. Among prokaryotes, both photosynthetic and nonphotosynthetic forms show gliding motility. Cyanobacteria and Chloroflexaceae are examples for the first group. The filamentous, sulfur-containing Beggiatoales (Pringsheim, 1949; Strohl and Larkin, 1978); the single-celled, sulfur-droplet-lacking Leucotrichales or short cell–chained *Vitreoscilla* group (Reichenbach and Golecki, 1975); the unicellular *Mycoplasma* group (Bredt and Radestock, 1977), and the class of Flexibacteriae (Soriano, 1947, 1973) with the orders of Cythophagales and Myxobacteriales are all examples of heterotrophic gliders. The obvious similarities between the photosynthetic and nonphotosynthetic forms induced most investigators to presume an analogous mechanism of movement, and it is for this reason that nonphotosynthetic organisms are also discussed. The taxonomic terms used here follow the ones proposed by Reichenbach and Dworkin (1981a). However, a number of gliding organisms are of uncertain taxonomic origin [e.g., the recently discovered genus *Oscillochloris* (Gorlenko and Pivovarova, 1977; Gorlenko and Korotkov, 1979)].

Among eukaryotic organisms, gliding occurs in pennate (raphe-possessing) diatoms (Gordon and Drum, 1970, Harper, 1980), in some Zygnematales [especially desmids (Pao-Zun Yeh and Gibor, 1970; Harper, 1980)], and in the monospores and spermatia of red algae (Nultsch, 1980). Gliding has also been described in Euglenophyceae (Günther, 1928; Jarosch, 1958; Suzaki and Williamson, 1985, 1986a,b,c,d), in other flagellate algae (Brokaw, 1962), and in protozoan Gregarines (Prell, 1921b; Schrevel et al., 1983).

Another point of interest is the diverse form of gliding organisms. Most of them glide as individual cells and the cell shape covers the range from the simple rodlike *Cytophaga* cell to the rigid silica-armored diatom frustule or the complex desmid assembled from two half cells. Also, the multicellular filamentous representatives show a wide variety of construction principles ranging from the caterpillarlike *Simonsiella* (Pangborn et al., 1977), the straight *Oscillatoria* (Schmid, 1921), the helical *Spirulina* (Van Eykelenburg, 1977, 1979), and the triradiate *Starria* (Fig. 1.2, Lang, 1977). In addition, some organisms (e.g., the Myxobacteria), undergo a complex metamorphosis during their cell cycle (Reichenbach, 1974): Starting as a single cell, Myxobacteria, such as *Chondromyces*, after aggregation develop into a complex treelike fruiting body (Fig. 1.3; Reichenbach and Dworkin, 1981a).

In the following sections, the specific characteristics of the most important groups of gliding organisms are discussed based on the available information and structured according to their systematic position.

Gliding Prokaryotes

All gliding prokaryotes possess a typical gram-negative cell wall structure (Glauert and Thornley, 1969; Drews and Weckesser, 1982). Different groups have peptidoglycan layers of variable thickness. Myxobacteria are covered by a more flexible coating, whereas Cyanobacteria have an extremely rigid murein sacculus. None of the gliding prokaryotes bear flagella at any time, nor are they capable of swimming. Flexing, bending, jerking, and twitching are common, especially among rod-shaped unicellular forms. In contrast, filamentous forms generally display continuous movements. From the vast multitude of gliding prokaryotes, only a few species have been investigated so far; nevertheless, the lack of a uniform principle of the locomotory mechanism is obvious.

Beggiatoales and Leucotrichales

Both orders include heterotrophs, which are often regarded as descendants of early cyanobacterial forms. This proposed relationship is based mainly on morphological similarities such as the presence of junctional pores in the cell wall, the rotation during forward locomotion [e.g., in *Beggiatoa* (Pringsheim, 1949; Drawert and Metzner-Küster, 1958)], the release of hormogonia in *Leucothrix* and *Thiothrix* (Reichenbach and Golecki, 1975), and the formation of necridia in *Beggiatoa* (Kolkwitz, 1909; Strohl and Larkin, 1978). Recent physiological studies (Smith, 1973; Cohen et al., 1975) also emphasize the close relationship. The linear velocity of forward propulsion ranges from 0.2 μm s^{-1} in *Vitreoscilla* (Pringsheim, 1951; Costerton et al., 1961; Doetsch and Hageage, 1968) to 8 μm s^{-1} in *Beggiatoa* sp. (Jarosch, 1958; Nelson and Castenholz, 1982). The obvious similarities to *Oscillatoria* induced most investigators to presume an analogous mechanism of movement (see following section on Cyanobacteria). Confirmatory observations, however, are still lacking.

Flexibacteriae

The Flexibacteriae comprise all the remaining heterotrophic bacterial gliders. It is therefore an artificial group, and any close phylogenetic relationship between the Myxobacteriales and the Cythophagales remains questionable. The first order is characterized by a unique life cycle in which special fruiting bodies are developed, whereas the Cytophagales comprise the heterogeneous remainder.

Measured speeds are up to 2.5 μm s^{-1} in *Cythophaga* (Stanier, 1942), but speeds are often lower. Colonies of spreading bacteria frequently produce circular or periodic patterns on agar surfaces. Recent research on gliding has concentrated on the Flexibacteriae, resulting in a number of models, hypotheses, and speculations.

Extensive studies of the cell surface of *Myxococcus xanthus* (MacRae and McCurdy, 1976; Kaiser, 1979; Dobson and McCurdy, 1979; Dobson et al., 1979; Kaiser and Crosby, 1983) and other Myxobacteria, Flexibacteria, and even Cyanobacteria (MacRae et al., 1977; Lounatmaa et al., 1980; Dick and Stewart, 1980; Lounatmaa and Vaara, 1980; Vaara, 1982; Vaara et al., 1984) have revealed polar fimbriae. These pili are supposed to move the cell, after making contact with a substrate, by alternately stretching and shortening. The absence of fimbriae in some motile strains of *Myxococcus* (Dobson et al., 1979) and the smooth (rather than crooked) tip of the pili are not compatible with this hypothesis. Besides, it is difficult to explain well-coordinated locomotion with such a mechanism.

Intracellular, periodic structures in some strains of *Myxobacteria* (Schmidt-Lorenz und Kühlwein, 1968; 1969), such as *Myxococcus xanthus* (Fig. 1.4; Burchard et al., 1977), have been suggested as being involved in the mechanism of movement. They consist of bundles of thin filaments shaped like herringbones that run just below the membrane originating from one cell pole. It has been proposed that a system similar to the actin–myosin complex undergoes contraction and relaxation to push cells forward in a manner similar to that of an inchworm. Recent proof of reversible inhibition of movement by proteolytic enzymes (Burchard et al., 1977) and the necessity of the presence of calcium (Womack et al., 1989) to induce and continue gliding indicate participating contractile proteins. At present, however, no light-microscopical observations of movement exist to confirm this hypothesis.

Similarly, the model by Lünsdorf and Reichenbach (1989) considered chainlike structures (strands) regarded as motor elements isolated from cell wall preparations of *Myxococcus fulvus*. The strands demonstrate a highly ordered periodicity of rings and connecting rows of longitudinal elements (Fig. 1.5). Contraction of the chain elements should cause undulating distortions of the cell envelope, resulting in movement. However, so far the whole arrangement has been demonstrated in only a single negatively stained cell. Furthermore, these topological elements could not be detected in freeze-fractured cells, so further confirmation will be necessary.

Recently, Koch (1990) suggested a more general concept for gliding motility based on sacculus contraction and expansion of cells powered by a chemiosmotic process (Mitchell, 1956; Pate, 1988; Khan, 1988). Charges (protons or other ions) are transported across the cytoplasmic membrane by either photosynthesis, respiratory activity, or ATPase action, resulting in an ion gradient between the interior and the outside of the cell (Koch, 1988). Shunts or channels allow the positive charges to return, and this is accompanied by osmotic shrinkage or expansion that, finally, is thought to propel the cell. However, up to now data are available only for the gram-negative sacculus of *Escherichia coli* (White et al., 1968; Johnson and White, 1972; Woldringh and Nanninga, 1985; Woeste, 1988), so that a generalization is difficult, especially since *E. coli* does not glide. In addition, there is another conceptional difficulty within this model: Many gliders exhibit continuous rotation during forward locomotion [e.g., *Flexibacter* or *Oscillatoria* (Burkholder, 1934; Ridgeway and Lewin, 1988)]. It is not easy to explain rotation by a simple sacculus contraction. Furthermore, many gliders are multicellular filamentous organisms, therefore, the contraction of each individual cell has to be coordinated with that of the others for a vectorial displacement.

There is a possible connection between this model and the goblet-or disk-shaped structures discovered on the surface of *Flexibacter* and *Cytophaga* species (Ridgway and Lewin, 1973; Ridgway et al., 1975). These organelles have been discussed as modified flagellar basal bodies that could function as submicroscopical wheels moving the cell along the substrate (Pate and Chang, 1979; Pate et al., 1979). But it is still unclear how steering and coordination of the sense of rotation in each hypothetical motor unit function. An alternative function of the goblets could be an involvement in secretion or in the chemiosmotic process.

The behavior of latex spheres attached to the surface of *Flexibacter polymorphus* and *Cytophaga* U67 suggests that the particles are adsorbed at binding sites (domains) in the outer membrane. In a video recording, these particles can be seen to be moved on helical tracks around the cells, which can continue for hours. In analogy, motility could be brought about by anchoring the cell to the substrate and displacing the anchor points relative to the cell. Even so, it is unclear how these anchor points are moved along

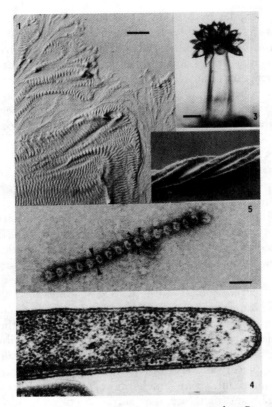

Figure 1.1. Swarming waves of *Stigmatella erecta* on an agar surface. Bar = 50 μm. [Courtesy of Dr. H. Reichenbach (from Reichenbach and Dworkin, 1981a) with permission of Springer-Verlag, New York.]

Figure 1.2. Scanning electron-microscopic photograph of a filament of the triradiated cyano-bacterium *Starria zimbabweënsis*. Bar = 10 μm. [Courtesy of Dr. N. J. Lang (from Lang, 1977) with permission of the Journal of Phycology.]

Figure 1.3. Mature fruiting body of *Chondromyces apiculatus* in situ on a filter paper. Bar = 100 μm. [Courtesy of Reichenbach (Reichenbach and Dworkin, 1981) with permission of Springer-Verlag, New York.]

Figure 1.4. Transmission electron micrographs of ultrathin sections of microfilaments in *Myxococcus xanthus* showing a herringbone arrangement. Bars = 200 nm. [Courtesy of Dr. C. A. Burchard (from Burchard et al., 1977) with permission of the American Society of Microbiology, Washington.]

Figure 1.5. Isolated negatively stained strands of *Myxococcus fulvus*. Small arrowheads indicate subunits of the rings and large arrowheads show the central elongated masses. Bar = 30 nm. (Courtesy of H. Lünsdorf.)

on the periphery of the cells (Lapidus and Berg, 1982). *Flexibacter* displays a movement accompanied by sinistral rotation with a constant torque similar to cyanobacteria.

As mentioned earlier, gliding is always accompanied by slime secretion (Kolkwitz, 1897; Fletcher and Floodgate, 1973; Marshall and Cruickshank, 1973). Therefore, a number of investigators assume that motility is intimately related to the release of slime (Humphrey et al., 1979; Keller et al., 1983; Burchard, 1986). Slimes are in most cases carbohydrates that wrap the cells with a sticky mucus layer and enable them to become attached to a suitable surface. Although it is indisputable that they facilitate tethering, it is difficult to explain movement only by slime secretion. Therefore, Keller et al. (1983) postulate a polar (located at both ends of a cell) and alternating slime extrusion. As a consequence, the surface tension will be lowered in the vicinity of the secretion site, causing an unbalanced force that propels the cell. Prerequisites for such a mechanism are secretion at a reasonable rate and a sufficient concentration difference between the two poles of the cell, both of which have not yet been demonstrated experimentally.

So far most work on gliding in this taxonomic group has concentrated on *Myxococcus xanthus*. Isolation and analysis of mutants to identify motility genes (Hodgkin and Kaiser 1977, 1979a,b; Kaiser, 1979; Kaiser and Crosby, 1983; Kaiser, 1984) has only partially elucidated the problem. The organisms utilize two genetically different systems; Gliding of single cells has been found to involve 22 gene loci, and social swarm gliding involves nine gene loci (Pate, 1988; McBride et al., 1989).

Chloroflexaceae

Only scattered data are available about the motility in Chloroflexaceae (Pierson and Castenholz, 1971, 1978). Up to now, *Chloroflexus aurantiacus* is the only species described on the basis of axenic cultures (Castenholz and Pierson, 1981). It is an orange to greenish filament of variable length, capable of slow movements (0.04 μm s^{-1}). The characteristic bacteriochlorophylls are contained in small *Chlorobium*-type vesicles (Pierson and Castenholz, 1978). In the 1970s two new genera were described: *Chloronema* (Dubinina and Gorlenko, 1975) and *Oscillochloris* (Gorlenko and Pivovarova, 1977). The mechanism of gliding in these organisms cannot be predicted, since the fine structure has not yet been investigated.

Cyanobacteria

Cyanobacteria (also called blue-green algae) strongly differ from all other eubacteria, in that they use two photosystems and a water-splitting reaction like higher plants. The photosynthetic pigments are located on the surface of thylakoids in highly efficient reaction centers and antenna assemblies called phycobilisomes, a situation similar to that in the red algae.

Occurrence and Phenomenology of Movement. Movements have been observed in numerous cyanobacteria, especially in benthic species of the Chroococcales and Hormogonales. Whereas single cells or colonies of the Chroococcales display jerking or slow, twitching movements (Desikachary, 1958; Pringsheim, 1968), the members of the Hormogonales exhibit well-coordinated gliding movements (Geitler, 1925). Nevertheless, the occurrence and nature of movement among the second group are very heterogeneous: The often constricted trichomes of the Nostocaceae glide, not unlike the movement of an earthworm without rotation (Harder, 1918). In some cases (e.g., most Scytonemataceae), movement is restricted to hormogonia, which are produced by either a specific liberation process or simple trichome fragmentation (Fritsch, 1945). In the Oscillatoriaceae, the trichomes glide with remarkably high velocities of up to 11 μm s^{-1} (Halfen and Castenholz, 1970), rotating around their longitudinal axis; the direction of rotation is specific for a given species (Correns, 1897). The velocity depends on temperature, light conditions, and the consistency of the substratum. The trichomes do not possess any obvious morphologic polarity and change their direction of movement frequently. The time intervals between these changes seem to be random, and trichomes have been observed to move for hours in a single direction (Nultsch, 1961). The family has been named because of the oscillations of the tip seen in two-dimensional projection under the microscope; they are brought about by the rotation of the asymmetrically crooked tip of the rotating filament (Kolkowitz, 1896). Gliding trichomes often describe slightly curved or circular paths when moving on agar or wet sand (Burkholder, 1934; Thomas, 1970). When describing a circle, the direction of revolution always depends on the sense of trichome rotation. Clockwise rotation (as in *Phormidium uncinatum*) drives the trichome in a right-handed circle, counterclockwise rotation (*Oscillatoria princeps* var. *maxima*) on a left-handed track. However, nonrotating species, such as *Nostoc* or *Anabaena*, can also display circular tracks (Lazaroff and Vishniac, 1964).

Trichomes sometimes show a helical coiling, when both tips of a trichome move toward each other. Since both ends rotate in opposite directions, the rotation twists the trichome. Starting with a single loop, the helical coil elongates until the whole trichome looks like a helix (Jarosch, 1963a). Reversal of movement unwinds the trichome. Similar helices have also been observed in other gliders [e.g., *Herpetosiphon giganteus* (Reichenbach and Dworkin, 1981b)].

Historical Perspectives. Since Adanson (1767), researchers have tried to elucidate the engima of gliding, and a number of hypotheses have been developed to explain motility in cyanobacteria, ranging from plausible models to wild speculations. Some hypotheses are based on changes of surface tension or electrokinetic processes, the propulsion by slime secretion, or the contraction or rotation of proteinaceous filaments. During the last three decades research has concentrated mainly on the latter of the two hypotheses, with a number of variations on the theme.

Contraction of filaments was first suggested by Schmid (1921, 1923), who interpreted moving light reflexes as contracting zones on the surface of trichomes. Stereoscopic pictures filmed by Ullrich (1926, 1929) were regarded as confirmation of Schmid's hypothesis until Schulz (1955) reinvestigated the observations and demonstrated the earlier results to be optical artifacts. In 1971, Halfen and Castenholz proposed a new model based on the theoretical considerations of Jarosch (1955, 1963b): They proposed that helically arranged contractile fibrillar elements located in one of the cell wall layers were responsible for locomotion in the Oscillatoriaceae, which rotate during forward locomotion; and that fibrillar elements parallel to the long axis of the trichome (which have not yet been shown in electron microscopical preparations) were responsible for locomotion in the Nostocaceae, which do not rotate during movement. Central to this hypothesis is that it is necessary for the filament to exert a force against the underlying substrate sufficient to propel itself forward.

The ejection and subsequent swelling of slime have for a long time been regarded as a possible basis of movement. Fechner (1915) tried to explain gliding by anisotropic swelling and presumed a preferential secretion at the terminal cells of trichomes. Schmid (1921), however, was able to demonstrate isotropic swelling and continuous secretion along the whole surface of the trichome. In addition, electron-microscopical studies confirmed that terminal cells of trichomes are often dead, or the cytoplasm content degenerated, and are consequently unable to produce slime (Lamont, 1969a; Shukovsky and Halfen, 1976).

Prell (1921a) formulated a modified hypothesis. He suggested helically arranged membrane pores, out of which the slime is squeezed in the form of small columns. Both the continuous ejection of new material and the swelling of the mucilage columns were suggested to propel the trichome. The helical arrangement of the pores was thought to induce a tangentially directed force and, consequently, rotation. However, such pores have yet to be found (Schulz, 1955; Metzner, 1955; Haxo et al., 1987), and slime columns have never been observed.

A number of investigations have employed ink or carmine particles that are adsorbed at the surface of the trichome (Correns, 1897; Hosoi, 1951; Schulz, 1955). During movement, these particles are moved along the length of the filament by the slime to which they stick; this has been taken as an indication that the slime plays an active role in motility. As the trichome moves forward, the mucilage moves backward, as visualized by the attached particles. When the trichome is tightly fixed in its position, the slime is moved along a helical track around the trichome. Changes of speed or reversals of movement are often accompanied by a change in the movement of adsorbed particles (Drews and Nultsch, 1962). The two preceding hypotheses are described later in more detail in conjunction with the pertinent structural elements.

Fine Structure of the Cell Envelopes. Prerequisite for the understanding of gliding is a thorough knowledge of the cellular architecture of gliders. The cell envelopes of gliders have therefore been widely studied. The term *cell envelope* is used here to describe all cell layers outside the cytoplasmic membrane. The obvious similarities between cyanobacteria and other gram-negative eubacteria (Allen, 1968; Butler and Allsopp, 1972, Drews and Weckesser, 1982) are consequently expressed in a similar terminology characterizing the individual layers in the envelope (Fig. 1.6). The former terminology (Jost, 1965) is given in brackets: The innermost layer of the complex cell wall—just outside of the cytoplasmic membrane—is the electron-dense peptidoglycan layer (L II layer). This "murein sacculus" is wider than that in most other bacteria, varying from 10 to 250 nm in *Oscillatoria princeps* (Halfen and Castenholz, 1971; Golecki, 1977; Jürgens et al., 1985). In most ultrathin sections, another layer (L I layer) of variable thickness is found between the membrane and the murein sacculus. This layer seems to be an artifact of fixation or embedding, since it is not present in freeze fracture (Golecki, 1977; Smarda et al., 1979) or cyro-ultrathin sections (Fig. 1.7; Hoiczyk and Häder, unpublished data). In most cyanobacteria, pores were found in the peptidoglycan layer, which can be classified into three different groups according to their location: (1) pores located in the cross walls (Metzner, 1955), which may be involved in signal transduction; (2) pores in the outer cell walls lining every cross wall on either side in circumferential rings (Frank et al., 1962; Haxo et al., 1987) (Fig. 1.8). In cross section these pores tilt inwards toward the cross walls and have an average diameter of 13 nm. They have been speculated to play a role in hormogonia liberation, trichome fragmentation (Lamont, 1969a), and slime secretion. A third type of pore is found in *Oscillatoria princeps* (Halfen and Castenholz, 1971; Shukovsky and Halfen, 1976) and *Starria zimbabweënsis* (Lang, 1977). These pores are irregularly distributed over the whole peptidoglycan layer. The individual pores have a diameter of about 70 nm and an average center-to-center spacing of 200 nm. It has been suggested that these "large pores" permit the transport of sheath precursor material and a steady supply of ATP to the postulated contractile fibrils in the outer membrane (Halfen and Castenholz, 1971). However, this type of pore might occur only in species

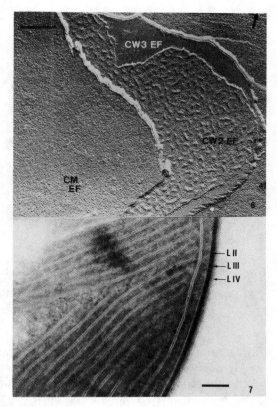

Figure 1.6. Exoplasmatic freeze-fractured face of *Anabaena variabilis* showing the cytoplasmic membrane (CM) and two outer fracture faces within the cell wall (CW2 and CW3). The arrow indicates the direction of shading. Bar = 200 nm. [Courtesy of Dr. J. R. Golecki (Drews and Weckesser, 1982) with permission of Blackwell Scientific Publications, Oxford.]

Figure 1.7. Electron-microscopical ultrathin cryosection of the cell wall in *Phormidium uncinatum* (strain isolated from Lake Baikal). L II to L IV indicate the cell wall layers (L I is seen only in sections of epon-embedded material). Bar = 0.1 µm (original photograph by Hoiczyk).

with a very rigid, thick peptidoglycan layer and facilitate diffusion in these cases.

Immediately distal to the murein layer is an electron transparent space (L III layer), which is probably also an artifact like the "L I layer," since it is not seen in freeze fracture or in isolated cell wall preparations (Drews and Weckesser, 1982). Outside the possibly artifactual L III layer there is a so-called outer membrane (L IV) with unit membrane–like structure. Frequently, cells are surrounded by additional layers. There is no uniform terminology for these external layers (Stanier and Cohen-Bazire, 1977; Cos-

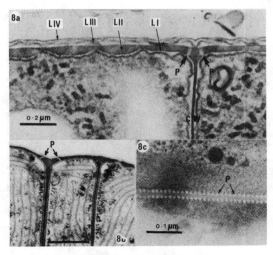

Figure 1.8. Junctional pores (P) in the outer cell walls on both sides of the cross walls (CW) in cross section in *Oscillatoria thiebautii* (a) and *Starria zimbabweënsis* (b; bar = 0.5 μm) and in top view in *Oscillatoria thiebautii* (c). L I to L IV indicate the four cell wall layers. [(a) and (c) courtesy of Dr. R. Lewin (Haxo et al., 1987) with permission of Blackwell Scientific Publications, Oxford and (b) courtesy of Dr. N. J. Lang (Lang, 1977) with permission of the Journal of Phycology.]

terton et al., 1978): Depending on the different appearance, the terms *sheath, slime, capsule,* or *glycocalyx* have been used. The thickness is up to 1000 nm, which is considerable compared to the size of the rest of the cell wall.

Most investigators have observed a more or less obvious fibrous network within the slime sheath (Ris and Singh, 1961; Pankratz and Bowen, 1963; Drews, 1973; Rippka et al., 1979; Schrader et al., 1982). In motile oscillatorean cyanobacteria these microfibrils are arranged roughly parallel to each other in a right-handed (e.g., *Oscillatoria chalybea*) or left-handed (e.g., *Lyngbya* spec.) helix. It is assumed that this helical orientation is a consequence of shear forces during rotation (Lamont, 1969b). Until now, it is not clear whether all motile cyanobacteria share such mucilaginous investment. Martin and Wyatt (1974) described amorphous shrouds in *Nostoc* that have no well-defined structure. In nonrotating or nonmotile species the arrangement of the microfibrils is either random or radiate (Frey-Wyssling and Stecher, 1954; Echlin, 1964; Tuffery, 1969). Dehydration during the embedding process might decrease the high-water content of these layers and irreversibly modify the fine structure.

So far structures have been described in only two layers, which might be involved in gliding: Van Eykelenburg (1977) has found a fibrillar array in the L I layer in *Spirulina platensis* and a similar pattern of microfibrils associated with the L III layer after shadowing. Freeze-fractured preparations of *Oscillatoria princeps* and grazing surface sections and replicas of *Oscillatoria animalis* and *Oscillatoria terebriformis* revealed analogous bands of fibrils running continuously from end to end of each trichome (Halfen and Castenholz, 1971; Halfen, 1973). The 5–8-nm-thick fibrils are highly organized in a left- or right-handed pattern around the trichome corresponding to the sense of rotation. Up to now, the limitations of the light microscope in studying living material have prevented proof of the interaction of the observed fibrils in gliding motility. However, as long as such putative motor elements are only visible in freeze fracture preparations and have been shown only rarely in ultrathin section, their actual existence, and thus their role in gliding, has to be interpreted with caution.

Mucilage Secretion. When they are observed in the light microscope, moving cyanobacterial trichomes glide with a steady motion, without apparent friction. The filaments move with an amazing force: When they are embedded in agar, *Phormidium uncinatum* trichomes move even through a very viscous medium, being stopped only by a concentration of 5 percent (Fig. 1.9). It is interesting that the velocity is highest at 0.4 percent and decreases in lower concentrations; this can be explained by the insufficient friction. Even at a concentration of 8 percent, the percentage of motile filaments is fairly high, even though the velocity is very low. The addition of a drop of India ink

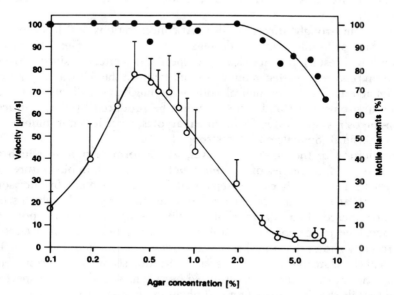

Figure 1.9. Velocity (ordinate, open circles with standard deviation) and percentage of motile trichomes (closed circles) of *Phormidium uncinatum* embedded in agar slabs of various concentrations (abscissa) (unpublished data by Häder).

suspension visualizes the mucilage, which is constantly secreted while the trichome moves (Niklitschek, 1934). In most oscillatoriacean cyanobacteria, the secreted slime forms a thin tube in which the filament glides. All cells in a filament contribute to slime production, so that the slime tube is thinnest at the front end and becomes thicker as the filament moves along. At the rear end, where the filament glides out of the tube, the slime tube collapses. The collapsed tube appears to be composed of parallel individual strands (Fig. 1.10), which are oriented at an angle of about 6° with respect to the long axis of the filament; this could be a result of rotation (Häder, 1987a). Since focusing through the empty slime tube does not show a layer of strands oriented in the opposite direction, another interpretation could be that the tube forms folds stained by the India ink particles on the surface that is not in contact with the substratum (Niklitschek, 1934). Despite the existence of different sets of pores in the peptidoglycan layer (see earlier), no experimental evidence has been found to suggest their role in slime extrusion.

Scanning electron microscopy after cryofixation of *Phormidium uncinatum* has not revealed any new insights into the extrusion process (Häder, 1987b). In other species, such as *Anabaena cylindrica* (Walsby, 1968), slime rings have been described and their independent movement has been considered as an indication for an autonomous gliding mechanism in each cell. Whatever the role of mucilage in forward propulsion may be, slime is undisputedly regarded as a lubricant for the adhesive surface, which facilitates movement.

Lysozyme strongly affects motility: At concentrations of 0.1 percent, the percentage of motile filaments decreases to almost zero (Fig. 1.11; Häder, unpublished result). It is not clear how specific this effect is, since lysozyme has been reported to affect a number of elements in the cell wall complex of cyanobacteria. The hypothetical contractile mechanism discussed for the generation of surface undulations cannot be produced by the contractile actin–myosin system often found in eukaryotes, since it is not inhibited by cytochalasin B (Spangle and Armstrong, 1973).

In addition to the outer slime layer, which forms a tube in which the filaments glide, trichomes of *Phormidium* form a sheath in old cultures (6–7 weeks). This sheath can be separated from the filament by ultrasonic treatment, after which it can be seen to form curls (Fig. 1.12), which seems to be composed of parallel rows of filaments. Whether or not the pores are artifacts due to ice formation still needs clarification. Scanning electron microscopy shows that the sheath tears along a helical line (Fig. 1.13). The physical correlate of the sheath as seen in ultrathin sections is not completely clear. The sheath cannot be hydrolyzed by enzymes, and infrared spectroscopy reveals that it consists of a carbohydrate with no protein components involved. Thin-layer chromatography of the monomers produced by acid hydrolysis show the presence of mannose, glucose, galactose, fucose, and

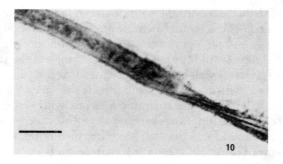

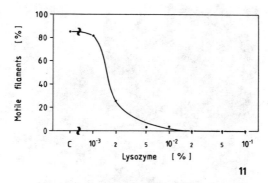

Figure 1.10. End of a trichome of *Phormidium uncinatum* leaving the slime tube that is visualized by India ink particles. Bar = 10 μm. (Original photography by Häder.)

Figure 1.11. Effect of lysozyme on the percentage of motile filaments in *Phormidium uncinatum*. (Häder, unpublished result.)

rhamnose, whereas in the slime material only glucose could be demonstrated. (Lack of sufficient material may, however, have been responsible for the failure to demonstrate the other sugar components.) Further differences are a helical fibrillar arrangement in the sheath, a characteristic birefringence, and a characteristic yellow-green autofluorescence of the sheath in blue excitation; none of which are observed in the outer slime tube.

Shear Mechanism. Central to the concept of shear mechanism is the idea that a filament must necessarily exert a force against the underlying substrate with sufficient strength to propel the trichome. Halfen and Castenholz (1971) suggested that right- or left-handed helically arranged protein fibrils undergo contractions that result in surface undulations. In contrast, Jarosch (1963b) assumes rotation of the fibrils results in surface undulations. In *Oscillatoria princeps*, Halfen and Castenholz (1971) found such fibrils with a diameter of 5–8 nm located adjacent to the peptidoglycan layer (L II layer, Fig. 1.14). The right-handed helical arrangement of pitch 60° corresponds to the helix described by any surface point during rotation and moving. The basis of such a motor system would be that reversible changes in the structure of the protein helices generate well-coordinated waves traveling along the trichome surface. The closely fitting elastic slime tube is the mechanical counterpart for these waves that push the trichome forward with respect to the slime sheath. Therefore, the amplitude of the waves must only be sufficient to exert the translational force through the outer membrane (L IV layer). The firm attachment of the mucilage to the substrate is a prerequisite for gliding.

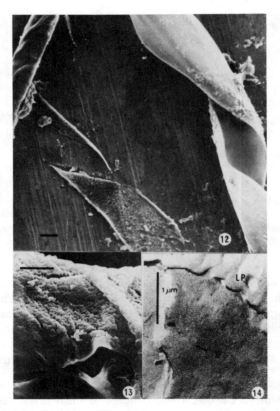

Figure 1.12. Curled sheath of *Phormidium uncinatum* separated from the trichome by ultrasonic treatment. Bar = 2 μm. (Original photograph by Hoiczyk.)

Figure 1.13. The sheath opens along a helical line during ultrasonic treatment of *Phormidium uncinatum*. Bar = 0.5 μm. (Original photograph by Hoiczyk.)

Figure 1.14. Fibrils adjacent to the L II cell wall layer in *Oscillatoria princeps* visualized as a platinum carbon replica from an isolated cell wall fragment. LP large pore, F fibrils in orientation represented by the arrow (Halfen, 1979). (With permission of Springer-Verlag, Heidelberg.)

If the trichome has no contact, the sheath is displaced over the surface—often in a helical manner.

We have attempted to visualize the postulated surface undulations by combining high-resolution light microscopy (Normarski interference contrast and 100 × oil immersion objective) and stroboscopic illumination with variable flash frequencies and have reasoned that the traveling waves have either a very small amplitude, an excessively fast frequency (beyond the flicker frequency of the human eye), or both (Häder, unpublished). However, the experiment failed to demonstrate visible traveling waves. Another attempt, using the newly developed acoustic microscope [ELSAM, Leitz, Wetzlar, FRG (Thaer et al., 1982)] also failed to prove the existence of surface undulations.

Shadowed surface replicas of *Oscillatoria terebriformis* and *Oscillatoria animalis* revealed raised areas, which were interpreted as clusters of fibrils frozen in contraction by the fixation process (Halfen, 1973). The height above the remaining surface of these areas was calculated to be 11 nm, which is regarded as sufficient to produce an effective contact with the surrounding slime tube.

The hypothesis that contractile filaments produce surface undulations that operate against the surrounding slime tube still needs confirmation and clarification:

1. It is not clear why most investigators have failed to observe fibrils incorporated into one of the cell wall layers, especially in ultrathin sections.
2. In nonrotating species, such as *Anabaena* or *Nostoc* hormogonia, potentially contractile filaments have never been observed (though postulated), a fact that could be interpreted by assuming that gliding in these species is based on a different mechanism.
3. It is not clear at all how trichome powers and controls the surface undulations that travel along the whole length of the trichome. An even bigger problem exists for coordinated reversals, such as those during photophobic reactions (Häder, 1974, 1976).
4. If the slime tube acts as a counterpart for surface undulations, the gelatinous material must have two almost contradictory characteristics: Optimal friction is necessary for transmission of the mechanical force, but for easy gliding a minimum of friction is needed.

The alternative hypothesis (namely, slime ejection and swelling) is even more difficult to explain than a model using surface undulations:

1. Slime ejection implies the existence of structures acting as jets. Up to now, only the junctional pores have been found to be widespread

among motile cyanobacteria. If these circumferential rows of pores are in fact the extrusion organelles of mucilage, then rotation is difficult to explain. They are arranged radially so they would have no effect on rotation. Besides, the junctional pores have not been found to extend through the whole cell wall; rather, they terminate just below the outer membrane. A simple jet-propulsion mechanism is therefore hardly credible. Furthermore, the process of actual slime secretion has never been observed in moving cyanobacteria.

2. Coordination and reversal of movement are problems similar to those in the alternative hypothesis. Since the trichomes move equally well in both directions, two sets of slime extrusion jets have to be postulated.

3. Calculations indicate that the amount of slime necessary to power the trichome by jet propulsion and subsequent swelling exceeds the total content of the cells within a short time (Holton and Freemann, 1965). In contrast, Schmid (1921) has calculated that the actual amount of slime secreted per hour in *Oscillatoria jenensis* is 1/212 of the trichome volume, thus sustaining movement for up to 8 days before the cell volume is used up.

4. Walsby (1968) pointed out that in *Anabaena cylindrica* mucilage rings moved over the cell surface, stopping at the constrictions between the cells. The rings never glided over specialized cells, such as heterocysts or akinetes, which are characterized by additional wall layers. He concluded that each cell has its own propulsion mechanism on the surface overlying the normal cell wall.

5. In some species, movement has been observed even when trichomes tunnel their way through agar with a concentration of up of 5 percent. The viscosity of the agar would require considerable power to propel the trichomes, which may be beyond the capabilities of a simple slime extrusion and swelling mechanism.

Although a number of details are known about motility in cyanobacteria, including fine structure, steering, possible motor elements, and energy supply, none of the presented models is sufficient to explain all aspects of gliding. This problem may be the consequence of insufficient knowledge, or it may reflect the fact that different groups utilize slightly or totally different mechanisms of gliding motility.

Control of Movement by External Factors. The movement of gliding cyanobacteria is controlled by a number of external factors; however, light seems to be the most important one (Häder, 1987a). Light controls motility by three different mechanisms: photokinesis, phototaxis, and photophobic responses.

Photokinesis describes the linear velocity of an organism dependent on the ambient fluence rate. When an organism moves faster at a given light intensity than in the dark control, this is defined as positive photokinesis. Most species studied to date show an optimal velocity at an irradiance of several thousand lx (Nultsch, 1974). The molecular mechanism involves the photosynthetic apparatus and can be simplified by the statement that, under favorable light conditions, more photosynthetic energy is produced to power motility. Both cyclic and noncyclic electron transport have been found to be involved in one or another cyanobacterial species (Nultsch, 1969; Nultsch and Jeeji-Bai, 1966).

The second mechanism by which cyanobacteria respond to light is photo-taxis, a directed movement with respect to the direction of the impinging light (Häder, 1979). Cyanobacteria master this task by two basically different mechanisms, exemplified by Oscillatoriaceae and Nostocaceae: Oscillatoria-ceae use a trial-and-error mechanism. Filaments change their direction of movement at regular or irregular time intervals (Haupt, 1977); those oriented more or less parallel to the light beam prolong their movement toward the light source and shorten the path in the opposite direction. By this procedure, the population gradually moves toward the light source (positive phototaxis).

Nostocaceae, in contrast, are the only prokaryotes known to be capable of a true steering mechanism (Nultsch et al., 1979): When they move in a lateral light beam, their front end turns in the direction of the light source (positive phototaxis at low fluence rates) or away from the light source (negative phototaxis). In these organisms, the mechanism of orientation seems to involve the intracellular production of singlet oxygen ($^1O_2$) (Nultsch et al., 1979). Nultsch and Schuchart (1985) have proposed a model to explain the molecular mechanism of phototactic orientation in *Anabaena* (Fig. 1.15): A hypothetical signal processor is assumed to detect the internal light gradient in laterally impinging light and causes the front tip to bend. At high fluence rates, the photosynthetic apparatus is thought to produce high concentrations of $^1O_2$, which is sensed by a sign reversal generator, which in turn controls the direction of bending with respect to the light direction.

The third type of light response is made up of photophobic responses,

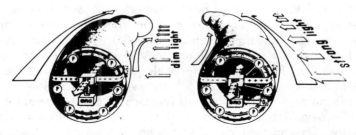

Figure 1.15. Model explaining phototactic orientation of *Anabaena variabilis* in lateral dim (a) and strong (b) light. Sp, signal processor; SRG, sign reversal generator. (Courtesy of Dr. W. Nultsch; with permission of Springer-Verlag, Heidelberg.) (From Nultsch and Schuchart, 1985.)

which can be defined as transient changes in the movement caused by sudden changes in the light intensity. The response can be a reversal of movement when the filament experiences a sudden decrease in the fluence rate or moves into a shaded (or very bright) area. In *Phormidium uncinatum*, the change in fluence rate is sensed by the photosynthetic pigments and detected in terms of a change in the linear electron transport through the plastoquinone pool (Häder, 1974). The result is a change in proton transport across the thylakoid membrane and thus a small change in the cytoplasmic electrical potential (Peschek et al., 1985). This mechanism seems to trigger Ca^{2+}-specific channels that open during a phobic response and allow a massive Ca^{2+} influx from the outside of the cell, thus amplifying the potential change (Häder and Poff, 1982). Extracellular and intracellular measurements of electrical potential changes indicate that the direction of movement is dictated by an electrical gradient between the front end and the rear end of the trichome (Häder, 1977; 1978a,b; 1987a). However, the mechanism by which the potential gradient is sensed and the means by which it controls the direction of surface undulations have not been revealed.

All these light responses lead to a consequential redistribution of the population in its habitat. Obviously, phototactic orientation will lead the organisms toward or away from the light source. The photophobic response prevents the trichomes from moving into the substratum or into too bright a light field. Likewise, photokinetic responses result in an oriented movement, since, for example, the filaments evacuate a certain area more rapidly with faster movement and increase the population density in this area when they move more slowly.

Gliding Eukaryotes

Diatoms

Almost all motile diatoms are pennate forms (in contrast to centric ones) and possess one or two raphe systems. Some centric forms, such as *Actinocyclus* and *Odontella*, seem to be the exception (Anderson et al., 1986; Pickett-Heaps et al., 1986). Depending on the arrangement of the raphe, the various forms glide on straight, slightly curved or circular paths (Nultsch, 1956). Rather than a uniform gliding, most organisms show an irregular motility, including frequent stops, reversals, jerks, or even spins. The linear velocity ranges from 0.2 to 25 μm s^{-1} (Drews and Nultsch, 1962; Harper, 1967), depending on light conditions (Nultsch, 1962), temperature (Hopkins, 1963), oxygen concentration (Hopkins, 1969), and endogenous factors, such as the phase in their circadian rhythm (Round and Palmer, 1966). Although some diatoms are among the fastest organisms known, the average velocity of such diatoms as *Hantzschia amphioxys* (Fauré-Fremiet, 1951) is clearly lower. As mentioned earlier, the existence of a raphe is essential for translocation. It consists of a slit in the rigid silicate frustule; however, it is not a simple cleft but has a complicated structure. In *Gyrosigma* (which may be regarded as a typical pennate diatom) the raphe runs in an undulated shape on both valves from one polar nodule to the central nodule and from there to the other polar nodule (Fig. 1.16). In cross section it is not a vertical cleft but consists of two (inner and outer) fissures oriented at an angle of about 45° to each other (Fig. 1.17). Near the polar nodule the inner fissure opens on the inner face of the cell wall in a funnel cleft. On both sides of the central nodule the outer and inner fissures are connected by vertical canals, and both inner fissures are joined by a horizontal canal.

Müller (1889) postulated that locomotion in diatoms is brought about by a movement of cytoplasm through the length of the outer fissure. This hypothesis is confirmed by particle displacement (attached to the outer cell wall) along the slits (Martens, 1940; Nultsch, 1962). Furthermore, cells are immobilized by plasmolysis during which the cytoplasm is withdrawn from

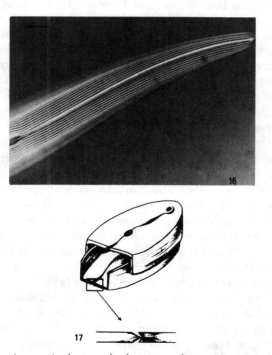

Figure 1.16. Light-microscopic photograph of a pennate diatom (*Gyrosigma*) after acid treatment to remove the living content of the cell showing the raphe system on one of the two valves. Bar = 10 μm. (Photograph by E. Hoiczyk.)

Figure 1.17. Schematic drawing of a pennate diatom, frustule with the center portion of the raphe in the valve enlarged in cross section. (Drawing by E. Hoiczyk.)

the raphe. In this sense the raphe can be regarded as a true locomotory "organelle." Even earlier, Ehrenberg (1838) based the first explanation on his observation of the extrusion of a slime trail that is left behind by a moving diatom. A rotating plasma stream was thought to run in the outer fissure and return in the inner fissure, resulting in a chain tread not unlike one used by a caterpillar or tank. The protoplasma was considered to generate shear forces that move the cell back and forth. This notion is also confirmed by the fact that the cells move only when the raphe is in contact with the surface; they do not move in a narrow glass tube, when only the outer edges of the valve touch the glass wall. Electron microscopial results, however, revealed the inconsistencies of the early explanations. In most cases the raphe is an open slit not covered by a membrane (Toman and Rosival, 1948; Reimann et al., 1965; Drum et al., 1966; Lauritis et al., 1967). Jarosch (1955, 1958, 1960) proposed rotating extracellular protein fibrils to be located in the outer fissure, which cause surface undulations (Jarosch, 1962). These submicroscopic waves were supposed to propel the cell relative to the mucilage adhering to the surface. Drum and Hopkins (1966) found a bundle of microtubules running in parallel through the raphe, which seems to refute the hypothesis. Instead, these authors proposed the existence of extensible polysaccharide fibrils to be responsible for propulsion. The observed direction of movement, however, is not compatible with this hypothesis. Another model (Harper and Harper, 1967) is based on osmosis, which is thought to reduce the internal pressure within the raphe and to allow the slime to move along easily. A lubricating liquid of unkown chemical consistency should support this process. A liquid also plays an important role in the model by Gordon and Drum (1970), who hypothesized that a liquid mucilage precursor lubricates the raphe. The steady dislocation is maintained by a capillary pressure difference between the two ends of the groove. None of the theories proposed up to now are capable without contradictions of satisfactorily explaining the complex movements. A recent model, however, proposes a theoretically reasonable mechanism for gliding in pennate diatoms: It suggests cooperative action between action bundles that, intracellularly, run parallel to the raphe and the extrusion of polysaccharide fibrils (Edgar and Pickett-Heaps, 1984).

A novel hypothesis was proposed by Gordon (1987). The chrysophycean alga *Ochromonas tuberculatus* possesses so-called discobolocysts, projectiles that are ejected at a speed of up to 260 m s^{-1} (as fast as a bullet from a small rifle). The source of energy is thought to be derived from hydration of one to three hydrogen bonds per monomer of the mucopolysaccharide that composes discobolocysts. The hydrogen bonding seems to induce a conformational change rather than a swelling of the polysaccharide. Theoretical considerations have shown that the degree of hydration of the mucopolysaccharides would result in a motive force sufficient also to propel pennate diatoms. The actin microfilament bundles inside the cell and parallel to the raphe could control the direction of movement, and the efficiency of the proposed motor could be as high as 99 percent.

Desmids

No other group of gliders exhibits a greater variety of shapes than the desmids (crescent moons, stars, extraterrestrial spheres). Their attractive appearance provoked very early systematic, microscopical research that revealed their ability to move (Ehrenberg, 1838; Nägeli, 1849; von Siebold, 1849). Most desmids are composed of two distinct halves, separated by a median constriction (sinus) and connected by an isthmus. Movement is generally slow; speeds up to 1 μm s^{-1} (Neuscheler, 1967) are average, and only rarely have higher velocities been reported (Klebs, 1885).

Gliding is always accompanied by the secretion of gelatinous slime material, which can be easily stained with India ink or dyes indicative of carbohydrates. Two types of locomotion can be distinguished according to the position of slime secretion. In the group characterized by *Closterium moniliferum*, both tips of the sickle-shaped cells alternately secrete mucilage. The resulting locomotion is a sequence of somersaults flipping the cell forward (Stahl, 1880). In the *Cosmarium* type, usually only the elder half-cell extrudes mucilage while moving, pushing the whole cell forward. The leading younger semicell is lifted above the surface, forming a variable angle with the substrate (Klebs, 1885). However, reversals of movement can be induced (e.g., by spatial or temporal changes in the irradiance). Under these circumstances the former leading half-cell starts producing a new slime stream while the rear stops secretion, reversing polarity and the direction of movement (Fig. 1.18, Häder and Wenderoth, 1977).

Both types are similar in that mucilage is secreted through a pore apparatus located at the terminal region of both half cells (Lütkemüller, 1902; Mix, 1966). The individual pores extend through the multilayered cell wall (Drawert and Metzner-Küster, 1961; Lott et al., 1972; Kiermayer and Staehelin, 1972; Mix, 1972, 1975; Gerrath, 1975). Although recent investigations (Chardard, 1977) shed some light on the fine structure of the pores and the ejected mucus, further investigations are required for a complete understanding.

In contrast to other gliding movements, the interpretation of the mechanism for motility in desmids is unanimous (Schröder, 1902; Kol, 1927; Jarosch, 1958; Pao-Zun Yeh and Gibor, 1970): Vectorial slime secretion and subsequent swelling of the ejected material allow the cell to be propelled like a jet.

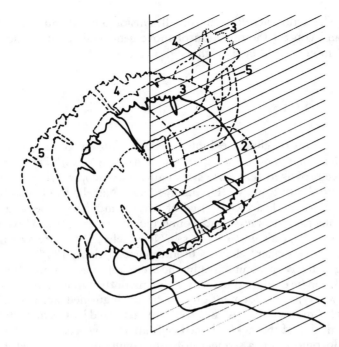

Figure 1.18. Reversal of movement in *Micrasterias truncata* by inducing a new slime secretion from the front end when moving into a dark area (shaded zone). Numbering indicates consecutive movement phases at 1-minute intervals. (After Häder and Wenderoth, 1977.)

Euglenoid Movements in Flagellates

In addition to flagellar movement, which is covered elsewhere in this volume (chapters 4 and 5), a number of flagellates have been observed to use gliding motility. Since several species of the genus *Euglena* and the related euglenoid *Astasia* show this type of locomotion fairly regularly under specific conditions, it is also termed euglenoid, metabolic, or peristaltic movement. The mechanics of this movement in *Euglena* are based on the pellicular strips, which constitute the outermost layer of the cell. These helically arranged strips seem to be passively moved by connecting fibrillar structures arranged underneath (Mikolajczyk and Kuznicki, 1981). Apparently, light governs not only flagellar movements (Häder et al., 1981; Häder, 1987c), but also euglenoid movements, since sudden changes in the fluence rate induce transient body contractions (Mikolajczyk and Diehn, 1976).

In *Euglena fusca*, each pellicular strip carries a row of particles, the relative position of which can be analyzed by video microscopy. During euglenoid movement, the strips slide past each other, which results in both oscillatory bending and rounding up of the cell. The maximal displacement between adjacent strips is 2.3 μm at a maximal speed of 0.4 μm s^{-1} (Suzaki and Williamson, 1985). No change was observed in the relative position of the particles in individual strips, which contradicts the hypothesis that movement is brought about by local contraction of the pellicular elements. The ultrastructure of the motile apparatus was analyzed in *Euglena ehrenbergii*, *E. oxyuris*, and *Astasia longa* (Suzaki and Williamson, 1986a,d). In cross section, individual pellicular strips are S-shaped, with the long ends hooking each element into the adjacent ones (Fig. 1.19). The whole arrangement is covered by the plasma membrane. Each strip carries periodic projections on the upward pointing edge, accompanied by four microtubules (two on each side) running parallel to the edge of the strip. It can be speculated that the microtubules function as rails that facilitate gliding along the adjacent elements. The microtubules are also thought to be the site of force generation and thus are speculated to glide actively relative to the adjacent strip. Neighboring strips are interconnected by an array of traversing filaments that are perpendicular to the long axis of the strips. The role of these filaments could be to restrict the amplitude of relative gliding between strips. Immediately underneath the filaments are tubular elements of the endoplasmic reticulum.

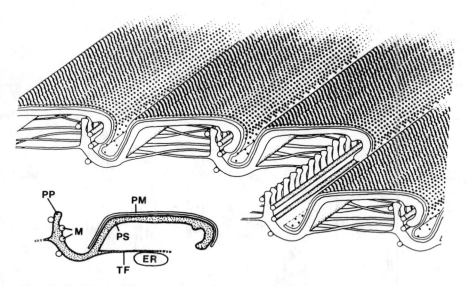

Figure 1.19. Schematic drawing of the pellicular strips (PS) underlying the plasma membrane (PM) in *Astasia longa*. The S-shaped individual elements are hooked into the adjacent ones by their edges. The upward-pointed edge carries a single row of periodic projections (PP) and four microtubules (M) running longitudinally, two on each side of the edge. Traversing filaments (TF) are positioned above endoplasmic reticulum tubuli (ER) and conect adjacent pellicular strips perpendicularly. [Courtesy of Dr. Suzaki (Suzaki and Williamson, 1986a), with permission of the Society for Protozoology, Lawrence.]

A computer simulation has shown that local and restricted sliding of adjacent pellicular strips is sufficient to explain all cellular shapes and types of movement observed in euglenoid motility (Fig. 1.20).

Cell models of *Astasia longa* were produced by treatment with detergents, which removed the flagellar membrane as well as all intracellular membranes and plasma membranes at groove regions of the cell surface. These cell models showed rounding up in the presence of calcium when the concentration was raised above 10^{-7} M. Addition of ATP strongly enhanced the reaction (Suzaki and Williamson, 1986c). The regulation by calcium in euglenoid movement was confirmed by Lonergan (1985), who also suggested the involvement of the actin–myosin system.

Some other flagellate algae, such as the green *Chlamydomonas*, also show gliding motility (for review, see Bloodgood, 1989). This movement, however, does not seem to be mediated by body contractions but rather by an alternative form of flagellar motility and will therefore not be discussed in detail in this chapter. This behavior is not powered by axonemal activity, since it is also found in mutants with paralyzed flagella (Bloodgood, 1977, 1981, 1988).

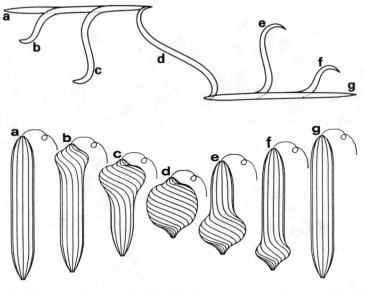

Figure 1.20. Localized and restricted sliding of pellicular strips with respect to the neighboring element are sufficient to explain rounding-up and bending movements observed in euglenoid motility as shown by a computer simulation. Top row changes in shape in a single pellicular strip, bottom row changes in cellular shape. [Courtesy of Dr. Suzaki (Suzaki and Williamson, 1986b).]

Ecological Significance of Gliding

Like other forms of motility, gliding movement opens up new possibilities for organisms to search actively for suitable habitats for survival and growth of the population. Motility is controlled and modulated by external stimuli, such as light, gravitational, chemical, or thermal gradients. These external factors are perceived by suitable receptors and, after appropriate amplification and sensory transduction, direct the search for a suitable environment. The result of such vectorial movements can be easily observed in gliders in their natural microhabitats. Some cyanobacteria prosper as mat-forming layers on the bottom of lakes, puddles, or even hot springs or pools. In this interwoven community, only constant movement directed by the available light intensity minimizes mutual shading. In addition, in some hot springs only adhesion to the bottom prevents the filaments from being washed away from the small range of temperatures that sustain growth and survival. Furthermore, gliding provides the organism with a mechanism to escape from lethal visible and ultraviolet irradiation during full sunshine (Häder, 1987b). Some *Oscillatoria* and other cyanobacteria species (as well as benthic diatoms) disappear from the surface under such light conditions (Nelson and Castenholz, 1982) and return at lower irradiances. This behavior is brought about by step-up photophobic responses, defined as a reversal of movement when an organism experiences a sudden increase in the fluence rate [e.g., when it moves from a dark area into a bright zone (Häder, 1987a)]. The opposite behavior, the step-down photophobic response, prevents the organisms from being buried in the sediment; a reversal of movement is induced when the organism moves from a moderately irradiated area into a dark field (Häder, 1987a,b).

The production of slime may be necessary not only for movement, but for attachment to the substratum. Trapping sediment particles strengthens the mud surface, and sometimes spectacular constructions like stromatolites can be formed.

The attachment to surfaces during movement may have been one of the important factors in the evolution of gliding, since this opens up the chance to spread and conquer niches that are not available for free-moving flagellate organisms, which are populated by gliding organisms. Slime is also considered as a means of trapping micronutrients in oligotrophic water, such as swamp ponds inhabited by desmids. In addition, slime may offer protection against viruses, lytic bacteria, or even protozoic predators. In eukaryotic forms, such as pennate diatoms, locomotion also confers a genetic advantage facilitating contact between individuals.

In conclusion, the extreme structural and physiological diversity found among gliding organisms argues against a uniform mechanism. For the same reason, gliding may not be a suitable characteristic to assemble different genera into the same taxonomic group. Certainly great efforts will be necessary to reveal the enigmas of motility in the different gliding organisms and to understand the structural basis, the physiological mechanisms, and the control and motor elements employed to facilitate smooth and coordinated movements in single and multicellular organisms.

References

Adanson, M. 1767. Un mouvement particulier découvet dans plante appellée tremella. Mem. Acad. Sci. Paris: 415–431.

Allen, M. M. 1968. Ultrastructure of the cell wall and cell division of unicellular green algae. *J. Bacteriol.* 96:842–852.

Anderson, R. A., Medlin, L. K., and Crawford, R. M. 1986. An investigation of the cell wall components of *Actinocyclus subtilis* (Bacillariophyceae). *J. Phycol.* 22:466–479.

Bean, B. 1984. Microbial geotaxis. In G. Colombetti and F. Lenci, (eds.), *Membranes and Sensory Transduction*. Plenum, New York, pp. 163–198.

Bloodgood, R. A. 1977. Motility occurring in association with the surface of the *Chlamydomonas* flagellum. *J. Cell Biol.* 75:983–989.

Bloodgood, R. A. 1981. Flagella-dependent gliding motility in *Chlamydomonas*. *Protoplasma* 106:183–192.

Bloodgood, R. A. 1988. Gliding motility and the dynamics of flagellar membrane glycoproteins in *Chlamydomonas reinhardtii*. *J. Protozool.* 35:552–558.

Bloodgood, R. A. 1989. Gliding motility and the flagellar surface. In R. A. Bloodgood, (ed.), *Ciliary and Flagellar Membranes*. Plenum Press, New York.

Bredt, W., and Radestock, U. 1977. Gliding motility of *Mycoplasma pulmonis*. *J. Bacteriol.* 130:937–938.

Brokaw, C. J. 1962. Flagella. In R. A. Lewin (ed.), *Physiology and Biochemistry of Algae*. Academic Press, New York, pp. 595–601.

Burchard, A. C., Burchard, R. P., and Kloetzel, J. A. 1977. Intracellular periodic structures in the gliding bacterium *Myxococcus xanthus*. *J. Bacteriol.* 132:666–672.

Burchard, R. P. 1981. Gliding motility. *Ann. Rev. Microbiol.* 35:497–529.

Burchard, R. P. 1986. The effect of surfactants on the motility and adhesion of gliding bacteria. *Arch. Microbiol.* 146:147–150.

Burkholder, P. R. 1934. Movement in Cyanophyceae. *Quart. Rev. Biol.* 9:438–459.

Butler, R. D., and Allsopp, A. 1972. Ultrastructural investigations in the Stigonemataceae. *Arch. Microbiol.* 82:283–299.

Castenholz, R. W. 1968. The behavior of *Oscillatoria terebriformis* in hot springs. *J. Phycol.* 4:132–139.

Castenholz, R. W., and Pierson, B. K. 1981. Isolation of members of the family Chloroflexaceae. In M. P. Starr, H. Stolp, H. G. Trüper, A. Balows, and H. G. Schlegel (eds.), *The Prokaryotes*. Springer-Verlag, Berlin, pp. 290–298.

Chardard, R. 1977. La secretion de mucilage chez quelques desmidiales I. Les pores. *Prostistologica* 13:241–251.

Cohen, Y., Padan, E., and Shilo, M. 1975. Facultative anoxygenic photosynthesis in the cyanobacterium *Oscillatoria limnetica J. Bacteriol.* 123:855–861.

Correns, C. 1897. Über die Membran und die Bewegung der Oscillatorien. *Ber. Dtsch. Bot. Ges.* 15:139–148.

Costerton, J. W., Geesey, G. G., and Cheng, K. J. 1978. How bacteria stick. *Sci. Am.* 238:86–95.

Costerton, J. W. F., Murray, R. G. E., and Robinow, C. F. 1961. Observations on the motility and the structure of *Vitreoscilla. Can. J. Microbiol.* 7:329–339.

Desikachary, T. V. 1959. Cyanophyta. *New Delhi Agricultural Research*, pp. 1–687. New Delhi, India.

Dick, H., Stewart, W. D. P. 1980. The occurrence of fimbriae on a N_2-fixing cyanobacterium which occurs in lichen symbiosis. *Arch. Microbiol.* 124:107–109.

Dobson, W. J., and McCurdy, H. D. 1979. The function of fimbriae in *Myxococcus xanthus*. I. Purification and properties of *M. xanthus* fimbriae. *Can. J. Microbiol.* 25:1152–1160.

Dobson, W. J., McCurdy, H. D., and MacRae, T. H. 1979. The function of fimbriae in *Myxococcus xanthus*. II. The role of fimbriae in cell-cell interactions. *Can. J. Microbiol.* 25:1359–1372.

Doetsch, R. N., and Hageage, G. J. 1968. Motility in prokaryotic organisms: Problems, points of view and perspectives. *Biol. Rev.* 43:317–362.

Drawert, H., and Metzner-Küster, I. 1958. Fluoreszenz- und elektronenmikroskopische Untersuchungen an *Beggiatoa alba* und *Thiothrix nivea. Arch. Microbiol.* 31:422–434.

Drawert, H., and Metzner-Küster, I. 1961. Licht und elektronenmikropische Untersuchungen an Desmidiaceen. I. Mitt. Zellwand und Gallertstrukturen bei einigen Arten. *Planta* 56:213–228.

Drews, G. 1959. Beiträge zur Kenntnis der phototaktischen Reaktionen der Cyanophyceen. *Arch. Protistenk.* 104:389–430.

Drews, G. 1973. Fine structure and chemical composition of the cell envelopes. In N. G. Carr and B. A. Whitton (eds.), *The Biology of Blue-Green Algae*. Blackwell, Oxford, pp. 96–116.

Drews, G., and Nultsch, W. 1962. Spezielle Bewegungsmechanismen bei Einzellern (Bakterien, Algen). In W. Ruhland (ed.), *Handbuch der Pflanzenphysiologie*, Vol. 17/2. Springer-Verlag, Berlin, pp. 876–919.

Drews, G., and Weckesser, J. 1982. Function, structure and composition of cell walls and external layers. In N. G. Carr, and B. A. Whitton, (eds.), *The Biology of Cyanobacteria*. Blackwell, London, pp. 333–358.

Drum, R. W., Hopkins, J. T. 1966. Diatom locomotion, an explanation. *Protoplasma* 62:1–33.

Drum, R. W., Pankratz, H. S., and Stoermer, E. F. 1966. Electron microscopy of diatom cells. In J. G. Helmcke, and W. Kreiger (eds.), *Diatomeenschalen im Electronenmikroskopischen Bild*, pp. 1–25 and 613 plates.

Dubinina, G. A., and Gorlenko, V. M. 1975. New filamentous photosynthetic green bacteria containing gas vacuoles. *Microbiology* 44:452–458.

Dworkin, M. 1983. Tactic behaviour of *Myxococcus xanthus. J. Bacteriol.* 154:452–459.

Echlin, P. 1964. The fine structure of the blue-green alga *Anacystis montana f. minor* grown in continuous illumination. *Protoplasma* 58:439–457.

Edgar, L. A., and Pickett-Heaps, J. D. 1984. Diatom locomotion. *Prog. Phycol. Res.* 3:47–88.

Ehrenberg, C. G. 1838. *Die Infusionsthierchen als vollkommene Organismen.* Leopold Voss, Leipzig.

Fauré-Fremiet, E. 1951. The tidal rhythm of the diatom *Hantzschia amphioxys. Biol. Bull. mar. biol. Lab. Woods Hole* 100:173–177.

Fechner, R. 1915. Die Chemotaxis der Oszillarien und ihre Bewegungserscheinungen überhaupt. *Z. Bot.* 7:289–364.

Fletcher, M., and Floodgate, G. D. 1973. An electron-microscopic demonstration of an acidic polysaccharide involved in the adhesion of a marine bacterium to solid surfaces. *J. Gen. Microbiol.* 74:325–334.

Frank, H., Lefort, M., and Martin, H. H. 1962. Elektronenoptische und chemische Untersuchungen an Zellwänden der Blaualge *Phormidium uncinatum. Z. Naturf. B.* 7:262–268.

Frey-Wyssling, A., and Stecher, E. 1954. Über den Feinbau des *Nostoc*-Schleimes. *Z. Zellforsch. Mikrosk. Anat.* 39:515–519.

Fritsch, F. E. 1945. *The Structure and Reproduction of the Algae,* Vol. II. Cambridge University Press, Cambridge.

Geitler, L. 1925. Synoptische Darstellung der Cyanophyceen in morphologischer und systematischer Hinsicht. *Beih. Bot. Centralbl.* 41:163–294.

Gerrath, J. F. 1975. Notes on desmids ultrastructure. I. Cell wall and zygote wall of *Cylindrocystis brebissonii.* II. The replicate division septum of *Bamubsina brebissonii. Beih. Nova Hedwigia* 42:103–113.

Glauert, A. M., and Thornley, M. J. 1969. The topography of the bacterial cell wall. *Ann. Rev. Microbiol.* 23:159–198.

Golecki, J. R. 1977. Studies on ultrastructure and composition of cell walls of the cyanobacterium *Anacystis nidulans. Arch. Microbiol.* 114:35–41.

Gordon, R. 1987. A retaliatory role for algal projectiles, with implications for the mechanochemistry of diatom gliding motility. *J. Theor. Biol.* 126:419–436.

Gordon, R., and Drum, R. W. 1970. A capillarity mechanism for diatom gliding locomotion. *Proc. Nat. Acad. Sci. U.S.A.* 67:338–344.

Gorlenko, V. M., and Korotkov, S. A. 1979. A new filamentous green bacteria with gas vacuoles, *Oscillochloris trichoides* nov. comb. In J. M. Nichols, (ed.), *Abstracts of the Third International Symposium on Photosynthetic Prokaryotes,* Oxford. University of Liverpool, England, p. A16.

Gorlenko, V. M., and Pivovarova, T. A. 1977. On the belonging of the blue-green alga, *Oscillatoria coerulescens* Gicklhorn 1921, to a new genus of Chlorobacteria, *Oscillochloris* nov. gen. *Izv. Akad. Nauk. SSSR, Ser. Biol.* 3:396–409.

Günther, F. 1928. Über den Bau und die Lebensweise der Euglenen, besonders der Arten *E. terricola, geniculata, proxima, sanguinea* und *luccus* nov. spec. *Arch. Protistenk.* 60:511–590.

Häder, D.-P. 1974. Participation of two photosystems in the photophobotaxis of *Phormidium uncinatum. Arch. Microbiol.* 96:255–266.

Häder, D.-P. 1976. Phobic reactions between two adjacent monochromatic light fields. *Z. Pflanzenphysiol.* 78:173–176.

Häder, D.-P. 1977. Influence of electric fields on photophobic reactions in blue-green algae. *Arch. Microbiol.* 114:83–86.

Häder, D.-P. 1978a. Extracellular and intracellular determination of light-induced potential changes during photophobic reactions in blue-green algae. *Arch. Microbiol.* 119:75–79.

Häder, D.-P. 1978b. Evidence of electrical potential changes in photophobically reacting blue–green algae. *Arch. Microbiol.* 118:115–119.

Häder, D.-P. 1979. Photomovement. In W. Haupt, and M. E. Feinleib, (eds.), *Ency-*

clopedia of Plant Physiology, New Series, Vol. 7, *Movement.* Springer-Verlag, Berlin, pp. 268–309.

Häder, D.-P. 1987a. Photomovement. In P. Fay and C. van Baalen, (eds.) *The Cyanobacteria.* Elsevier, New York, pp. 325–345.

Häder, D.-P. 1987b. Photosensory behavior in procaryotes. *Microbiol. Rev.* 51:1–21.

Häder, D.-P. 1987c. Polarotaxis, gravitaxis and vertical phototaxis in the green flagellate, *Euglena gracilis. Arch. Microbiol.* 147:179–183.

Häder, D.-P., Colombetti, G., Lenci, F., and Quaglia, M. 1981. Phototaxis in the flagellates, *Euglena gracilis* and *Ochromonas danica. Arch. Microbiol.* 130:78–82.

Häder, D.-P., and Poff, K. L. 1982. Dependence of the photophobic response of the blue-green alga, *Phormidium uncinatum,* on cations. *Arch. Microbiol.* 132:345–348.

Häder, D.-P., and Wenderoth, K. 1977. Role of three basic light reactions in photo-movement of desmids. *Planta* 137:207–214.

Halfen, L. N. 1973. Gliding motility of *Oscillatoria*: Ultrastructural and chemical characterization of the fibrillar layer. *J. Phycol.* 9:248–253.

Halfen, L. N. 1979. Gliding movements. In W. Haupt and M. E. Feinleib, (eds.), *Encyclopedia of Plant Physiology* Vol. 7, *Physiology of Movements.* Springer-Verlag, Berlin, pp. 250–267.

Halfen, L. N., and Castenholz, R. W. 1970. Gliding in a blue-green alga: A possible mechanism. *Nature* 225:1163–1165.

Halfen, L. N., and Castenholz, R. W. 1971. Gliding motility in the blue-green alga *Oscillatoria princeps. J. Phycol.* 7:133–145.

Harder, R. 1918. Über die Bewegung der Nostocaceen. *Z. Bot.* 10:177–244.

Harper, M. A. 1967. Locomotion of diatoms and "clumping" of blue-green algae. Ph.D. Thesis. University of Bristol, pp. 1–171.

Harper, M. A. 1976. The migration rhythm of the benthic diatom *Pinnularia viridis* on pad silt. *New Zeal. J. Mar. Freshw. Res.* 100:381–384.

Harper, M. A. 1980. Movements. In D. Werner, (ed.), *The Biology of Diatoms.* Blackwell, Oxford, pp. 224–249.

Harper, M. A., and Harper, J. F. 1967. Measurements of diatom adhesion and their relationship with movement. *Br. Phycol. Bull.* 3:195–207.

Haupt, W. 1977. *Bewegungsphysiologie der Pflanzen.* Georg Thieme Verlag, Stuttgart.

Haupt, W. 1983. Photoperception and photomovement. *Philos. Trans. R. Soc. Lond. Ser. B* 303:467–478.

Haxo, F. T., Lewin, R. A., Lee, K. W., and Li, M.-R. 1987. Fine structure of *Oscillatoria* (Trichodesmium) aff. *thiebautii* (Cyanophyta) in culture. *Phycologia* 26:433–456.

Henrichsen, J. 1972. Bacterial surface translocation: A survey and a classification. *Bact. Rev.* 36:478–503.

Ho, J., and McCurdy, H. D. 1979. Demonstration of positive chemotaxis of cyclic GMP and 5-AMP in *Myxococcus xanthus* by means of a simple apparatus for generating practically stable concentration gradients. *Can. J. Microbiol.* 25:1214–1218.

Hodgkin, J., and Kaiser, D. 1977. Cell-to-cell stimulation of movement in nonmotile mutants of *Myxococcus. Proc. Nat. Acad. Sci. U.S.A.* 74:2938–2942.

Hodgkin, J., Kaiser, D. 1979a. Genetics of gliding motility in *Myxococcus xanthus* (Myxobacterales): Genes controlling movements of single cells. *Mol. Gen. Genet.* 171:167–176.

Hodgkin, J., and Kaiser, D. 1979b. Genetics of gliding motility in *Myxococcus xanthus* (Myxobacterales): Two gene systems control movement. *Mol. Gen. Genet.* 171:177–191.

Holton, R. W., and Freemann, A. W. 1965. Some theoretical and experimental considerations of the gliding movement of blue-green algae. *Am. J. Bot.* 52:640.

Hopkins, J. T. 1963. A study of diatoms of the Ouse estuary, Sussex. I. The movement of the mud-flat diatoms in response to some chemical and physical changes. *J. Mar. Biol. Assoc. U.K.* 43:653–663.

Hopkins, J. T. 1966. Some light induced changes in behaviour and cytology of an estuarine mud-flat diatom. In *Light as an Ecological Factor. Symp. Br. Ecol. Soc.* 5:335–358, Blackwell, Oxford.

Hopkins, J. T. 1969. Diatom motility: Its mechanism, and diatom behaviour patterns in estuarine mud. Ph.D. Thesis, University of London, pp. 1–251.

Hosoi, A. 1951. Secretion of the slime substance in *Oscillatoria* in relation to its movement. *Bot. Mag.* 64:14–16.

Humphrey, B. A., Dickson, M. R., and Marshall, K. C. 1979. Physicochemical and in situ observations on the adhesion of gliding bacteria to surfaces. *Arch. Microbiol.* 120:231–238.

Jarosch, R. 1955. Untersuchungen über Plasmaströmungen. Doctoral Thesis, University of Vienna.

Jarosch, R. 1958. Zur Gleitbewegung der niederen Organismen. *Protoplasma* 50:277–289.

Jarosch, R. 1960. Die Dynamik im Characeen-Protoplasma. *Phyton* 15:43–66.

Jarosch, R. 1962. Gliding. In R. A. Lewin (ed.), *Physiology and Biochemistry of Algae*. Academic Press, New York, pp. 573–581.

Jarosch, R. 1963a. Gleitbewegungen und Torsionen von Oscillatorien. *Österr. Bot. Z.* 111:476–481.

Jarosch, R. 1963b. Grundlagen einer Schraubenmechanik des Protoplasmas. *Protoplasma* 57:448–500.

Johnson, R. Y., and White, D. 1972. Myxospore formation in *Myxococcus xanthus*. Chemical changes in the cell wall during cellular morphogenesis. *J. Bacteriol.* 112:849–855.

Jost, M. 1965. Die Ultrastruktur von *Oscillatoria rubescens* D. C. *Arch. Mikrobiol.* 50:211–245.

Jürgens, U. J., Golecki, R. J., and Weckesser, J. 1985. Characterization of the cell wall of the unicellular cyanobacterium *Synechocystis* PCC 6714. *Arch. Microbiol.* 142:168–174.

Kaiser, D. 1979. Social gliding is correlated with the presence of pili in *Myxococcus xanthus*. *Proc. Natl. Acad. Sci. U.S.A.* 74:5952–5956.

Kaiser, D. 1984. Genetics of Myxobacteria. In R. Rosenberg (ed.), *Myxobacteria: Development and cell interactions*. Springer-Verlag, New York, pp. 163–184.

Kaiser, D., and Crosby, C. 1983. Cell movement and its coordination in swarms of *Myxococcus xanthus*. *Cell Motil.* 3:227–245.

Kaiser, D., Manoil, C., and Dworkin, M. 1979. Myxobacteria: Cell interactions, genetics and development. *Ann. Rev. Microbiol.* 33:595–639.

Keller, K. H., Grady, M., and Dworkin, M. 1983. Surface tension gradients: Feasible model for gliding motility in *Myxococcus xanthus*. *J. Bacteriol.* 155:1358–1366.

Khan, S. 1988. Analysis of bacterial flagellar rotation. *Cell Motil. Cytoskel.* 10:38–46.

Kiermayer, O., and Staehelin, L. A. 1972. Feinstruktur von Zellwand und Plasmamembran bei *Micrasterias denticulata* nach Gefrierätzung. *Protoplasma.* 74:227–237.

Klebs, G. 1885. Über Bewegung und Schleimbildung bei den Desmidiaceen. *Biol. Zentralbl.* 5:353–367.

Koch, A. L. 1988. Biophysics of bacteria walls viewed as stress-bearing fabric. *Microbiol. Rev.* 52:337–357.

Koch, A. L. 1990. The sacculus contraction/expansion model for gliding motility. *J. Theor. Biol.* 142:95–112.

Kol, E. 1927. Über die Bewegung mit Schleimbildung einiger Desmidiaceen aus der hohen Tatra. *Fol. Krypt.* 1:435–442.

Kolkwitz, R. 1896. Über die Krümmungen bei den Oscillatorien. *Ber. Dtsch. Bot. Ges.* 14:422–431.

Kolkwitz, R. 1897. Über die Krümmungen und den Membranbau bei einigen Spaltalgen. *Ber. Dtsch. Bot. Ges.* 15:460–467.

Kolkwitz, R. 1915. Schizomycetes. Spaltpilze (Bacteria). In Botanischer Verein der Provinz Brandenburg (ed.), *Kryptogamenflora der Mark Brandenburg.* vol. 5. Gebrüder Borntraeger, Leipzig.

Lamont, H. C. 1969a. Sacrifical cell death and trichome breakage in an oscillatoriacean blue-green alga: The role of murein. *Arch. Mikrobiol.* 69:237–259.

Lamont, H. C. 1969b. Shear-oriented microfibrils in the mucilaginous investments of two motile oscillatoriacean blue-green algae. *J. Bacteriol.* 97:350–361.

Lang, J. N. 1977. *Starria zimbabweënsis* (Cyanophyceae) gen. nov. et sp. nov.: A filament triradiate in transverse section. *J. Phycol.* 13:288–296.

Lapidus, R. I., and Berg, H. C. 1982. Gliding motility of *Cytophaga* sp. strain U67. *J. Bacteriol.* 151:384–398.

Lauritis, J. A., Hemmingsen, B. B., and Volcani, B. E. 1967. Propagation of *Hantzschia* sp. Grunow daughter cells by *Nitzschia alba* Lewin & Lewin. *J. Phycol.* 3:236–237.

Lazaroff, N., and Vishniac, W. 1964. The relationship of cellular differentiation to colonial morphogenesis of the blue-green alga *Nostoc muscorum. J. Gen. Microbiol.* 35:447–457.

Lonergan, T. A. 1985. Regulation of cell shape in *Euglena gracilis.* VI. Localization of actin, myosin and calmodulin. *J. Cell Sci.* 77:197–208.

Lott, L. N. A., Harris, G. P., and Turner, C. D. 1972. The cell wall of *Cosmarium botrytis. J. Phycol.* 8:232–236.

Lounatmaa, K., and Vaara, T. 1980. Freeze-fracturing of the cell envelope of the *Synechocystis* CB3. *FEMS Microbiol. Lett.* 9:203–209.

Lounatmaa, K., Vaara, T., Österlund, K., and Vaara, M. 1980. Ultrastructure of the cell wall of a *Synechocystis* strain. *Can. J. Microbiol.* 26:204–298.

Lünsdorf, H., and Reichenbach, H. 1989. Ultrastructural details of the apparatus of gliding motility of *Myxococcus fulvus* (Myxobacterales). *J. Gen. Microbiol.* 135:1633–1641.

Lütkemüller, J. 1902. Die Zellmembran der Desmidiaceen. *Beitr. Biol. Pfl.* 8:347–414.

McBride, M. J., Weinberg, R. A., and Zusman, D. R. 1989. "Frizzy" aggregation genes of the gliding bacterium *Myxococcus xanthus* show sequence similarities to the chemotaxis genes of enteric bacteria. *Proc. Natl. Acad. Sci. U.S.A.* 86:424–428.

MacRae, T. H., Dobson, W. J., and McCurdy, H. D. 1977. Fimbriation in gliding bacteria. *Can. J. Microbiol.* 23:1096–1108.

MacRae, T. H., and McCurdy, H. D. 1976. Evidence for motility-related fimbriae in the gliding microorganism *Myxococcus xanthus. Can. J. Microbiol.* 22:1589–1593.

Marshall, K. C., and Cruickshank, R. H. 1973. Cell surface hydrophobicity and the orientation of certain bacteria at interfaces. *Arch. Mikrobiol.* 91:29–40.

Martens, P. 1940. La locomotion des Diatomees. *Cellule* 48:277–306.

Martin, T. C., and Wyatt, J. T. 1974. Extracellular investments in blue-green algae with particular emphasis on the genus *Nostoc*. *J. Phycol.* 10:204–210.

Metzner, J. 1955. Zur Chemie und zum submikroskopischen Aufbau der Zellwände, Scheiden und Gallerten der Cyanophyceen. *Arch. Mikrobiol.* 22:45–77.

Mikolajczyk, E., and Diehn, B. 1976. Light-induced body movement of *Euglena gracilis* coupled to flagellar photophobic responses by mechanical stimulation. *J. Protozool.* 23:144–147.

Mikolajczyk, E., and Kuznicki, L. 1981. Body contraction and ultrastructure of *Euglena*. *Acta Protozool.* 20:1–24.

Mitchell, P. 1956. Hypothetical thermokinetic and electrokinetic mechanisms of locomotion in micro-organisms. *Proc. R. Phys. Soc.* 25:32–34.

Mix, M. 1966. Licht- und elektronenmikroskopische Untersuchungen an Desmidiaceen. XII. Zur Feinstruktur der Zellwände und Mikrofibrillen einiger Desmidiaceen vom *Cosmarium*—Typ. *Arch. Mikrobiol.* 55:116–133.

Mix, M. 1972. Die Feinstruktur der Zellwände bei Mesotaeniaceae und Gonatozygaceae mit einer vergleichenden Betrachtung der verschiedenen Wandtypen der Conjugatophyceae und über deren systematischen Wert. *Arch. Mikrobiol.* 81:197–220.

Mix, M. 1975. Die Feinstruktur der Zellwände der Conjugaten und ihre systematische Bedeutung. *Beih. Nova Hedwigia* 42:179–194.

Müller, O. 1889. Durchbrechungen in der Zellwand und ihre Beziehungen zur Ortsbewegung der Bacillariaceen. *Ber. Dtsch. Bot. Ges.* 7:169–180.

Nägeli, C. 1849. *Gattungen einzelliger Algen, physiologisch und systematisch bearbeitet.* Zürich.

Nelson, D. C., and Castenholz, R. W. 1982. Light responses of *Beggiatoa*. *Arch. Microbiol.* 131:146–155.

Neuscheler, W. 1967. Bewegungs- und Orientierungsweise bei *Micrasterias denticulata* im Licht I. Zur Bewegungs- und Orientierungsweise. *Z. Pfl. Physiol.* 57:46–59.

Niklitschek, A. 1934. Das Problem der Oscillatorienbewegung. *Beih. Bot. Zbl, Abt. A* 52:205–254.

Nultsch, W. 1956. Studien über die Phototaxis der Diatomeen. *Arch. Protistenk.* 101:1–68.

Nultsch, W. 1961. Der Einfluß des Lichtes auf die Bewegung der Cyanophyceen. 1. Phototopotaxis von *Phormidium autumnale*. *Planta* 57:632–647.

Nultsch, W. 1962. Über das Bewegungsverhalten der Diatomeen. *Planta* 58:22–30.

Nultsch, W. 1969. Effect of desaspidin and DCMU on photokinesis of blue-green algae. *Photochem. Photobiol.* 10:119–123.

Nultsch, W. 1971. Phototactic and photokinetic action spectra of the diatom *Nitzschia communis*. *Photochem. Photobiol.* 14:705–712.

Nultsch, W. 1974. Der Einfluß des Lichtes auf die Bewegung phototropher Mikroorganismen. I. Photokinesis. *Abh. Marburg. Gelehrt. Ges.* 2:143–213.

Nultsch, W. 1980. Movements. In W. D. P. Stewart (ed.), *Algal Physiology and Biochemistry*. Blackwell, Oxford, pp. 864–893.

Nultsch, W., and Häder, D.-P. 1979. Photomovement of motile microorganisms. *Photochem. Photobiol.* 29:423–437.

Nultsch, W., and Jeeji-Bai. 1966. Untersuchungen über den Einfluß von Photosynthesehemmstoffen auf das phototaktische und photokinetische Reaktionsverhalten blaugrüner Algen. *Z. Pflanzenphysiol.* 54:84–98.

Nultsch, W., and Schuchart, H. 1985. A model of the phototactic reaction chain of the cyanobacterium *Anabaena variabilis*. *Arch. Microbiol.* 142:180–184.

Nultsch, W., Schuchart, H., and Höhl, M. 1979. Investigations on the phototactic orientation of *Anabaena variabilis*. *Arch. Microbiol.* 142:85–91.

Nultsch, W., and Wenderoth, K. 1973. Phototaktische Untersuchungen an einzelnen Zellen von *Navicula peregrina* (Ehrenberg) Kützing. *Arch. Mikrobiol.* 90:47–58.

Pangborn, J., Kuhn, D. A., and Woods, J. R. 1977. Dorsal-ventral differentiation in *Simonsiella* and other aspects of its morphology and ultrastructure. *Arch. Microbiol.* 113:197–204.

Pankratz, H. S., and Bowen, C. C. 1963. Cytology of blue-green algae. I. The cells of *Symploca muscorum*. *Am. J. Bot.* 50:387–399.

Pao-Zun Y., Gibor, A. 1970. Growth patterns and motility of *Spirogyra* sp. and *Closterium acerosum*. *J. Phycol.* 6:44–48.

Pate, J. L. 1988. Gliding motility in prokaryotic cells. *Can. J. Microbiol.* 34:459–466.

Pate, J. L., and Chang, L.-Y.E. 1979. Evidence that gliding motility in prokaryotic cells is driven by rotary assemblies in the cell envelopes. *Curr. Microbiol.* 2:59–64.

Pate, J. L., Petzold, S. J., and Chang. L. E. 1979. Phages for the gliding bacterium *Cytophaga johnsonae* that affect only motile cells. *Curr. Microbiol.* 2:257–262.

Peschek, G. A., Czerny, T., Schmetterer, G., and Nitschmann, W. H. 1985. Transmembrane proton electrochemical gradients in dark aerobic and anaerobic cells of the cyanobacterium (blue-green alga) *Anacystis nidulans*. Evidence for respiratory energy transduction in the plasma membrane. *Plant Physiol.* 79:278–284.

Pickett-Heaps, J. D., Hill, D. R. A., and Wetherbee, R. 1986. Cellular movement in the centric diatom *Odontella sinensis*. *J. Phycol.* 22:334–339.

Pierson, B. K., and Castenholz, R. W. 1971. Bacteriochlorophylls in gliding filamentous prokaryotes from hot-springs. *Nature* 233:25–27.

Pierson, B. K., and Castenholz, R. W. 1978. Photosynthetic apparatus and cell membranes of the green bacteria. In R. K. Clayton and W. R. Sistrom (eds.), *The Photosynthetic Bacteria*. Plenum, London, pp. 179–197.

Prell, H. 1921a. Zur Theorie der sekretorischen Ortsbewegung. I. Die Bewegung der Cyanophyceen. *Arch. Protistenk.* 42:99–156.

Prell, H. 1921b. Zur Theorie der sekretorischen Ortsbewegung. II. Die Bewegung der Gregarinen. *Arch. Protistenk.* 42:157–175.

Pringsheim, E. G. 1949. The relationship between bacteria and myxophyceae. *Bacteriol. Rev.* 13:47–98.

Pringsheim, E. G. 1951. The Vitreoscillaceae: A family of colourless, gliding filamentous organisms. *J. Gen. Microbiol.* 5:124–149.

Pringsheim, E. G. 1968. Cyanophyceen-Studien. *Arch. Mikrobiol.* 63:331–355.

Reichenbach, H. 1974. Die Biologie der Myxobakterien. *Biologie in unserer Zeit* 4:33–45.

Reichenbach, H., and Dworkin, M. 1981a. Introduction to the gliding bacteria. In M. P. Starr, H. Stolp, H. G. Trüper, A. Balows and H. G. Schlegel (eds.), *The Prokaryotes*. Springer-Verlag, Berlin, pp. 315–327.

Reichenbach, H., and Dworkin M. 1981b. The order Cytophagales (with Addenda on the genera *Herpetosiphon, Saprospira*, and *Flexithrix*). In M. P. Starr, H. Stolp, H. G. Trüper, A. Balows, and H. G. Schlegel, (eds.), *The Prokaryotes*. Springer-Verlag, Berlin, pp. 356–379.

Reichenbach, H., and Golecki, J. R. 1975. The fine structure of *Herpetosiphon*, and a note on the taxonomy of the genus. *Arch. Microbiol.* 102:281–291.

Reimann, B. E. F., Lewin, J. C., and Volcani, B. E. 1965. Studies on the biochemistry and fine structure of silica shell formation in diatoms. I. The structure of the cell wall of Cylindrotheca fusiformis. J. Cell Biol. 24:39–55.

Reimers, H. 1928. Über die Thermotaxis niederer Organismen. Jahrb. Wiss. Bot. 67:242–290.

Ridgway, H. F., and Lewin, R. A. 1973. Goblet-shaped sub-units from the wall of a marine gliding microbe. J. Gen. Microbiol. 79:119–128.

Ridgway, H. F., and Lewin, R. A. 1988. Characterization of gliding motility in Flexibacter polymorphus. Cell Motil. Cytoskel. 11:46–63.

Ridgway, H. F., Wagner, R. M., Dawsey, W. T., and Lewin, R. A. 1975. Fine structure of the cell envelope layers of Flexibacter polymorphus. Can. J. Microbiol. 21:1733–1750.

Rippka, R., Deruelles, J., Wazterbury, J. B., Herdmann, M., and Stanier, R. Y. 1979. Generic assignments, strain histories and properties of pure cultures of cyanobacteria. J. Gen. Microbiol. 111:1–61.

Ris, H., and Singh, R. N. 1961. Electron microscope studies on blue-green algae. J. Biophys. Biochem. Cytol. 9:63–79.

Round, F. E., and Palmer, D. J. 1966. Persistent vertical-migration rhythms in benthic microflora. II. Field and laboratory studies on the diatoms from the banks of the river Avon. J. mar. biol. Assoc. U.K. 46:191–214.

Schmid, G. 1918. Zur Kenntnis der Oszillarienbewegung. Flora 111:327–379.

Schmid, G. 1921. Über Organisation und Schleimausbildung bei Oscillatoria jenensis und das Bewegungsverhalten künstlicher Teilstücke. Jahrb. Wiss. Bot. 62:572–625.

Schmid, G., 1923. Das Reizverhalten künstlicher Teillstücke, die Kontraktilität und das osmotische Verhalten der Oscillatoria jenensis. Jahrb. Wiss. Bot. 62:328–419.

Schmidt-Lorenz, W., and Kühlwein, H. 1968. Intracelluläre Bewegungsorganellen der Myxobakterien. Arch. Mikrobiol. 60:95–98.

Schmidt-Lorenz, W. and Kühlwein, H. 1969. Beiträge zur Kenntnis der Myxobakterienzelle. Arch Mikrobiol. 68:405–426.

Schrader, M., Drews, G., Weckesser, J., and Mayer, H. 1982. Polysaccharide containing 6-o-methyl-D-mannose in Chlorogloeopsis PCC 6912. J. Gen. Microbiol. 128:273–277.

Schrevel, J., Caigneaux, E., Cros, D., and Philippe, M. 1983. The three cortical membranes of the gregarines. I. Ultrastructural Organisation of Gregarina blaberae. J. Cell Sci. 61:151–179.

Schröder, B. 1902. Untersuchungen über Gallertbildungen bei Algen. Verh. Naturhist. Med. Verein Heidelberg, N.F. 7:139–196.

Schulz, G. 1955. Bewegungsstudien sowie elektronenmikroskopische Membranuntersuchungen an Cyanophyceen. Arch. Mikrobiol. 21:335–370.

Shukovsky, E. S., and Halfen, L. N. 1976. Cellular differentiation of terminal regions of trichomes of Oscillatoria princeps (Cyanophyceae). J. Phycol. 12:336–342.

Smarda, J., Cáslaská, J., and Komárek, J. 1979. Cell wall structure of Synechocystis aquatilis (Cyanophyceae). Arch. Hydrobiol. Suppl. 56:154–165.

Smith, A. J. 1973. Synthesis of metabolic intermediates. In N. G. Carr and B. A. Whitton (eds.), The Biology of Blue-Green Algae. Blackwell, London, pp. 1–38.

Soriano, S. 1947. The Flexibacteriales and their systematic position. Antonie van Leeuwenhoek J. Microbiol. Serol. 12:215–222.

Soriano, S. 1973. Flexibacteria. Ann. Rev. Microbiol. 27:155–170.

Spangle, L., and Armstrong, P. B. 1973. Gliding motility of algae is unaffected by cytochalasin B. Exp. Cell Res. 80:490–493.

Stahl, E. 1880. Über den Einfluß von Richtung und Stärke der Beleuchtung auf einige Bewegungserscheinungen im Pflanzenreich. *Bot. Ztg.* 38:297–413.

Stanier, R. Y., and Cohen-Bazire, G. 1977. Phototrophic prokaryotes: The cyanobacteria. *Ann. Rev. Microbiol.* 31:225–274.

Stanier, R. Y. 1942. The Cytophaga group: A contribution on the biology of Myxobacteria. *Bacteriol. Rev.* 6:143–196.

Strohl, W. R., and Larkin, J. M. 1978. Cell division and trichome breakage in *Beggiatoa. Curr. Microbiol.* 1:151–155.

Suzaki, T., and Williamson, R. E. 1985. Euglenoid movement in *Euglena fusca*: Evidence for sliding between pellicular strips. *Protoplasma* 124:137–146.

Suzaki, T., and Williamson, R. E. 1986a. Ultrastructure and sliding of pellicular structures during euglenoid movement in *Astasia longa* Pringsheim (Sarcomastigophora, Euglenida). *J. Protozool.* 33:179–184.

Suzaki, T., and Williamson, R. E. 1986b. Cell surface displacement during euglenoid movement and its computer stimulation. *Cell Motil. Cytoskel.* 6:186–192.

Suzaki, T., and Williamson, R. E. 1986c. Reactivation of euglenoid movement and flagellar beating in detergent-extracted cells of *Astasia longa*: Different mechanisms of force generation are involved. *J. Cell Sci.* 80:75–89.

Suzaki, T., and Williamson, R. E. 1986d. Pellicular ultrastructure and euglenoid movement in *Euglena ehrenbergii* Klebs and *Euglena oxyuris* Schmarda. *J. Protozool.* 33:165–171.

Thaer, A., Hoppe, M., and Patzelt, W. J. 1982. Akustomikroskop Elsam. *Leitz-Mitt. Wiss. Tech.* 8:61–67.

Thomas, E. A. 1970. Beobachtungen über das Wandern von *Phormidium autumnale. Schweiz. Z. Hydrol.* 32:523–531.

Toman, M., and Rosival, M. 1948. The structure of the raphe of Nitzschiae. *Stud. Bot. Cechosl.* 9:26–29.

Tuffery, A. A. 1969. Light and electron microscopy of the sheath of a blue-green alga. *J. Gen. Microbiol.* 57:41–50.

Ullrich, H. 1926. Über die Bewegung von *Beggiatoa mirabilis* und *Oscillatoria jenensis*. I. *Planta* 2:295–324.

Ullrich, H. 1929. Über die Bewegungen der Beggiatoaceen und Oscillatoriaceen. II. *Planta* 9:144–194.

Vaara, T. 1982. The outermost surface structures of chroococcacean cyanobacteria. *Can. J. Microbiol.* 28:929–941.

Vaara, T., Ranta, H., Loutnamaa, K., and Korhonen, T. K. 1984. Isolation and characterisation of pili (fimbriae) from *Synechocystis* CB3. *FEMS Microbiol. Lett.* 21:329–334.

Van Eykelenburg, C. 1977. On the morphology and ultrastructure of the cell wall of *Spirulina plantensis. Antonie van Leeuwenhoek* 43:89–99.

Van Eykelenburg, C. 1979. The ultrastructure of *Spirulina platensis* in relation to temperature and light intensity. *Antonie van Leeuwenhoek* 45:369–390.

Verworn, M. 1889. *Psychophysiologische Protistenstudien.* Verlag Gustav Fischer, Jena.

Von Siebold, C. T. 1849. Über einzellige Pflanzen und Tiere. *Z. wiss. Zool.* 1:93–102, 270–294.

Wagner, J. 1934. Beiträge zur Kenntnis von *Nitzschia putrida* Benecke insbesondere ihrer Bewegung. *Arch. Protistenk.* 82:86–113.

Walsby, A. E. 1968. Mucilage secretion and the movements of blue-green algae. *Protoplasma* 65:223–238.

White, D., Dworkin, M., and Tipper, D. J. 1968. Peptidoglycan of *Myxococcus xanthus*: Structure and relation to morphogenesis. *J. Bacteriol.* 95:2186–2197.

Williams, R. B. 1965 Unusual motility of tube-dwelling pennate diatoms. *J. Phycol.* 1:145–146

Woeste, S. 1988. The mechanical properties of the isolated sacculi of *Eschericha coli*. Ph.D. dissertation, Indiana University.

Woldringh, C. L. and Nanninga, N. 1985. Structure of nucleoid and cytoplasm in the intact cell. In N. Nanninga (ed.), *Molecular Cytology of Escherichia coli* L. Academic Press, New York, pp. 161–197.

Womack, B. J., Gilmore, D. F., and White, D. 1989. Calcium requirement for gliding motility in Myxobacteria. *J. Bacteriol.* 171:6093–6096.

Chapter 2

Algal Chloroplast Movements

Gottfried Wagner and Franz Grolig

Organelle movement is a ubiquitous phenomenon of eukaryotic cells with an intact cytoskeleton. Apart from cell shape, cell development, and cell differentiation, the cytoskeletal array plays a key role in cellular and subcellular motility. Cytoplasmic movement is the major precondition to guarantee metabolic homeostasis; in fact, the larger the cell, the more extensive is the intracellular movement that is observed. Cytoplasmic movement is readily observed in plant cells, since the photosynthetic organelles are often transported either to optimize photosynthetic yield or to distribute photosynthetic metabolites. This review focuses on the movements of algal chloroplasts.

Among the plants that have been thoroughly investigated in this respect are several algae that may serve as model systems in studying the physiology of intracellular movements. In such research, algae offer several advantages over higher plants: (1) many algae are organized as if physiologically unicellular organisms, and the studies are less complicated than those in higher plant tissue; (2) most of the algae are aquatic, a feature that eases design of experimental protocols; (3) algal interphase cells are often haploid and therefore less cumbersome in application of the techniques of molecular genetics. On the other hand, of course, the results obtained in lower plants must be critically evaluated if they are to be extrapolated to other organisms, such as higher plants.

The two main types of intracellular movement are easily distinguishable from each other, although they may overlap or even be interconnected. The first type is called *cytoplasmic streaming*, a term that denotes the more or less continuous motion of a smaller or a larger fraction of the cytoplasm. The individual movement of numerous organelles on the whole causes an overall streaming of the cytoplasm, with the overall pattern of organelle distribution remaining unchanged. In the most coherent type of cytoplasmic streaming (*rotation* or *cyclosis*), the velocity distribution of the participating organelles is narrow.

The classical example is the internodal cell of characean algae (namely, *Chara* and *Nitella*). Characean endoplasm streams inside a cylinder of stationary cortical cytoplasm in which the chloroplasts form helically wound files paralleling the direction of endoplasmic streaming. Cytoplasmic streaming in Characean algae is described in more detail in Chapter 3 of this volume. In other algae, the streaming in opposite directions is not confined to two halves of the cortical cytoplasm; instead, longitudinal tracks with streaming in opposite direction may alternate. This *multistriate* streaming will be described here, inter alia, in the algae *Vaucheria* and *Acetabularia*.

Besides these fundamental types, several other modifications of cytoplasmic streaming show less regularity (e.g., *circulation* and *agitation*). However, these specificities of cytoplasmic streaming will not be discussed here and the reader is referred to the reviews [e.g., Haupt (1983a) and Kamiya (1986)].

In contrast to the more or less indefinite cytoplasmic streaming, which does not change the statistical distribution of cell organelles, there is a more advanced, organelle-specific type of intracellular movement: *the orientational movement of chloroplasts*. Like any of the sensory transduction chains in nature, at least three major components are needed to describe these orientational movements in sequential order: (1) a receptor to perceive the signal; (2) a sensory transducer converting the signal into a chemical message; (3) an effector to respond according to the information perceived. Often orientational chloroplast movement is a photoresponse. Thus, upon perception of the photosignal proper, chloroplasts begin to be rearranged in the new position, with or without respect to the direction of the impinging light (scalar information vs. vectorial information). For the latter type of response, the term *intracellular taxis* was used in the old literature to characterize these light-oriented intracellular movements. In fact, the phenotype shows some similarity to oriented movements of free-living organisms.

As a general rule, chloroplast photo-orientational movement occurs in two or three distinct and well-defined patterns. In light of low or medium intensity (e.g., white light of 1000–5000 lux or blue light with an irradiance of 1–10 W m^{-2}), chloroplasts are arranged in a way that allows maximum absorption of the impinging light. At high intensity (e.g., full sunshine), however, they are arranged in a pattern permitting minimum absorption. An operational terminology, which relates primarily to the light conditions, will describe the chloroplasts in "low-irradiance" or in "high-irradiance" position. Examples are the orientational chloroplast movements in the green algae *Mougeotia* and *Mesotaenium*, which will be described here in detail. In addition, some plants have a distinct dark arrangement of chloroplasts, which usually is a random distribution throughout the cytoplasm. Such a relaxation of light-triggered, but not light-oriented, chloroplast movement is seen, for example, in *Biddulphia*, *Eremosphaera*, and *Striatella*.

The Chloroplast as a Participant in Organelle and Cytoplasmic Streaming

Vaucheria

In the coenocytic alga *Vaucheria*, with its large central vacuole and peripheral multinucleate cytoplasm, the chloroplast movement is part of the cytoplasmic streaming, both in darkness and in light (Fig. 2.1). In total light, the chloroplast arrangements deviating from random distribution are understood in terms of optical physics (Senn, 1908), when light passes the reflective algal cylinder (see Fig. 2.5). No experimental information, however, is available on the sensory transduction chain here. This is in contrast to another phenomenon whereby, upon local irradiation of the coenocytic alga (see Fig. 2.1), cortical fibers can be seen to reticulate, and concomitantly, chloroplasts aggregate in this region. Local light therefore acts as a trap to the passing chloroplasts (Blatt et al., 1980). Several lines of evidence indicate that the cortical fibers enclose actin filaments and filament bundles (Fig. 2.2). Cytochalasin B (CB) was found both to inhibit streaming and to alter the organization of the cortical fibers, whereas treatment of the cells with phalloidin, which stabilizes the F-actin, protects the organelle movement from the action of CB. When they are decorated with a preparation of purified subfragment S1 from rabbit muscle myosin, the filament bundles appear fuzzy. In single filaments the distinct arrowhead pattern characteristic of S1-bound F-actin is clearly visible (Blatt et al., 1980). The organelles may move along the cortical actin fibers in much the same way as that proposed for characean algae (see chapter 3), but light apparently disrupts the cable assembly. Thus, the organelles are trapped in the irradiated region where the actin–myosin interaction is interrupted.

The light effect, as characterized by an action spectrum proper, is shown in Figures 2.3 and 2.4. The most striking feature is the very sharp action maximum at 473 nm. Sensitivity at this wavelength, compared to the neighboring wavelengths at 454 nm and 431 nm light, is up to an order of magnitude higher than comparable wavelength sensitivities of classic blue-

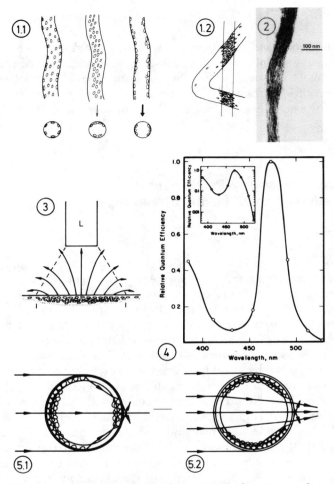

Figure 2.1. Chloroplast arrangements in *Vaucheria*. (1.1) Surface views and cross sections through a filament in darkness (left) and in unilateral light of low irradiance (center) and of high irradiance (right). Arrows above the cross sections denote the light direction of different irradiance. In the surface views, the light direction is normal to the plane of the paper. (1.2) Unilateral light of low irradiance, projected through a slit onto part of a bent filament; chloroplasts accumulate in the irradiated regions. (After Haupt, 1983a.)

Figure 2.2. Actin from *Vaucheria*. High-magnification electron micrograph of a negatively stained fiber filament that shows the manner in which bundles of 5–7-nm-diameter filaments join to form a larger cable. (From Blatt et al., 1980; with permission.)

Figure 2.3. Outward current from *Vaucheria* upon partial irradiation. Blue light is given through an optical fiber, L, producing a microbeam, the aperture of which is indicated by the dashed lines. The two bars underneath the accumulated chloroplasts denote the light–dark boundary. The arrows represent direction and strength of the current as measured by a vibrating electrode. (After Blatt et al., 1981.)

Figure 2.4. Action spectrum for chloroplast aggregation in *Vaucheria* at the 50 percent response level. For further details, see Hartmann (1983). *Inset:* Action spectrum plotted logarithmically. (From Blatt, 1983; with permission.)

Figure 2.5. Chloroplast arrangement in *Vaucheria*. Light refraction and chloroplast arrangement in low-irradiance (5.1) and in high-irradiance (5.2) mode in a filament, surrounded for demonstration by air. (After Senn, 1908.)

light ("cryptochrome") responses, such as the phototropic curvature of *Avena* (Thimann and Curry, 1961) and *Phycomyces* (Delbrück and Shropshire, 1960). On the other hand, at wavelengths below 430 nm the action spectrum is characteristic of the flavin absorption spectrum, which is supposed to be the chromophore of "cryptochrome." Thus, if a single sensory pigment is responsible for the chloroplast aggregation in *Vaucheria*, its identity remains a mystery.

Locally illuminated areas of *Vaucheria* reveal an electrical outward current (see Fig. 2.3), which results from an efflux of protons (Blatt et al., 1981) and generates a local hyperpolarization. This proton flux shows the same fluence response dependency and the same spectral sensitivity as the chloroplast aggregation, and the proton efflux slightly precedes chloroplast aggregation. Thus, directly or indirectly, the light-induced proton efflux may entail the observed chloroplast stop response. Two major areas remain open: (1) no information is available on how light absorption by the unidentified blue-light receptor affects proton transport through the cell membrane; (2) the mechanism of actin inactivation as a consequence of proton efflux is unclear.

It is of some interest to compare chloroplast aggregation with three growth-related photoresponses in *Vaucheria* (i.e., phototropic bending, apical expansion, and branching). Growth in *Vaucheria* is limited to the filament tip, and phototropic bending results from shifting of the growth center to the illuminated side of the cell apex (Kataoka, 1975b). Likewise, apical expansion (Kataoka, 1981) and branching (Kataoka, 1975b) reflect the enlargement of existing growth centers and the generation of new growth centers, respectively. All three growth-related photoresponses are mediated by blue light, and the former two at least occur over a time scale comparable to that for chloroplast aggregation (Kataoka, 1975a,b; 1981). Chloroplast aggregation, phototropic bending, apical expansion, and branching may be related mechanistically. Tip growth is generally associated with filament-based transport that delivers vesicles, filled with wall-precursor material, to the growth zone for deposition (Franke et al., 1972). In *Vaucheria*, such vesicles move along cytoplasmic fibers composed of actin (Blatt et al., 1980) and accumulate, along with mitochondria and dictyosomes, in a "hyaline cap" at the growing tip. At least superficially, this network appears identical to the actin fiber reticulum that forms preceding aggregation (Blatt and Briggs 1980). Thus, as Blatt (1983) suggested by analogy with aggregation, the light-induced changes in growth pattern could depend on similar alterations in the underlying actin bundle organization. However, long-term observations (Kicherer, 1985; Kataoka and Weisenseel, 1988) indicate that an additional factor must be involved. The blue-light–induced local current efflux starts to decay 3–4 minutes after the beginning of blue-light (BL) irradiation. At about 60 minutes, when the accumulation of chloroplasts is accomplished, the outward current is replaced by a gradually increasing inward current. Since the growth-related photoresponses in *Vaucheria* were always observed at sites of current entry, the onset of inward current apparently corresponds to the development of a new growth center at the BL-irradiated site.

Dichotomosiphon

Chloroplasts and other cytoplasmic particles in the fresh-water coenocytic green alga *Dichotomosiphon tuberulus* continuously stream together with the cytosol along the longitudinal axis of the algal body, in a multistriated and bidirectional fashion (Maekawa and Nagai, 1988). In response to light irradiation, the cytoplasm is moved toward the apical region and accumulates there; in darkness the cytoplasm migrates back to the basal region. Local accumulation of the cytoplasm, including chloroplasts, was induced when the alga was locally irradiated with blue light (Fig. 2.6). These movements were inhibited in the presence of colchicine (Maekawa et al., 1986). By immunofluorescence and electron microscopy it was revealed that the reorganization of microtubule bundles in the endoplasm along the length of the alga is directly correlated to the translocation of the cytoplasm. In the dark the classes of small bundles (1–15 microtubules) and of medium bundles (16–30 microtubules) were evenly distributed along the whole length of the specimens, with no bundles of more than 31 microtubules to be seen (Maekawa and Nagai, 1988). Under total light irradiation, the number of small bundles decreased considerably and medium bundles disappeared in the apical region, whereas in the basal region the number of medium bundles increased and large bundles were found as well (Fig. 2.7). Similarly, decomposition of bundles was observed where the coenocytium was locally irradiated. Hence, reversible formation and decomposition of the microtubule bundles presumably play a crucial role in the light-induced translocation of the cytoplasm in this alga. Bridgelike structures were often seen linking individual microtubules within bundles. These structures are tentatively supposed to be microtubule-associated proteins (MAPs). They are suspected to play an important role in the formation and decomposition of microtubule bundles in this alga. Preliminary observations show that in *Dichotomosiphon*, organelle accumulation upon local irradiation is most sensitive to blue light and far less sensitive to green or red light. However, an action spectrum as discussed earlier for *Vaucheria* (see Fig. 2.4) is not yet available.

Bryopsis, Caulerpa, and *Halimeda*

It is well known that in several siphonaceous algae (e.g., species of *Bryopsis, Caulerpa,* and *Halimeda*), parts of the thallus change color diurnally, alternating between dark green during the day and pale green or even colorless

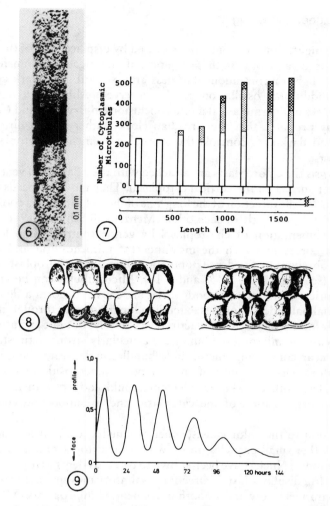

Figure 2.6. Chloroplast arrangement in response to unilateral light, projected through a slit onto part of *Dichotomosiphon*. Chloroplasts accumulate in dense package in the irradiated region (cf. Fig. 2.1.2). (From Maekawa et al., 1986.)

Figure 2.7. Typical distribution pattern of microtubules in *Dichotomosiphon*. Serial cross sections were obtained from the apical region to the basal region of the specimen (schematized in the abscissa). Each arrow corresponds to the position of a cross section examined. Open, dotted, and meshed columns represent the number of microtubules, making small, medium, and large bundles defined in the text. (After Maekawa and Nagai, 1988.)

Figure 2.8. Vertical section through a thallus of *Ulva*, with the chloroplast in each cell in day (left) and in night position (right), respectively. Synonyms in use are the following: day position = face position = position at periclinal wall; night position = profile position = position at anticlinal wall. (From Senn, 1908; with permission.)

Figure 2.9. Chloroplast movement in *Ulva* in continuous light of low irradiance (0.05 W m^{-2}) during several days. Changes in transmittance of the thallus, given on the ordinate in relative units, indicate chloroplast movement toward face and profile position, respectively, corresponding to day and night movement in the natural light–dark cycle. (After Britz et al., 1976.)

45

during the night. This phenomenon is caused by displacement of the bulk of chloroplasts in and out of these regions. If the algae are transferred to continuous light or continuous darkness after several light–dark cycles, at least one additional oscillation can be observed. Although a rather fast damping does not permit any definite conclusions in *Bryopsis* and *Caulerpa*, it is highly probable that the circadian clock, rather than a direct effect of light and darkness, controls the diurnal movement of cytoplasm and chloroplasts.

In *Bryopsis* the chloroplasts are attached with their flattened, ventral sides to the inner surface of the cortical cytoplasm. They move at speeds of up to 50 μm min^{-1} in the direction of the long axis of the cell, either in a coordinated fashion or independently of each other (Menzel and Schliwa, 1986a). Intracellular sedimentation of chloroplasts by gentle centrifugation leaves an intact cell cortex in which the movement of mitochondria and nuclei— normally obscured by the dense population of moving chloroplasts—can be observed (see Figs. 2.10, 2.15, and 2.16). Microtubules can be visualized using a variety of tubulin antibodies and are shown to form a dense, two-dimensional palisade of bundles extending longitudinally through the cortical cytoplasm. Parallel arrays of actin fibers closely, but not exclusively, co-localize with the microtubule bundles. Particularly strong actin staining is observed near converging microtubule bundles underneath the tip regions of the chloroplasts. Because of the extensive superposition of actin and microtubules, both cytoskeletal elements could cooperate in maintaining the spatial organization of the cytoplasm and in supporting chloroplast movement.

In addition to the chloroplasts, other organelles (e.g., nuclei, mitochondria, and other small particles) move within the cytoplasm in an organelle-specific manner: Nuclei have been shown to rotate as they move forward, and mitochondria display a fast, bidirectional saltatory movement, in contrast to the comparatively slow and smooth movement of chloroplasts (Menzel and Schliwa, 1986a). By taking advantage of the property of cytoplasmic exudates of *Bryopsis* to survive in vitro for a limited period of time (Fig. 2.11), it can be demonstrated that saltations of mitochondria and other particles in the cytoplasmic exudates follow exactly the course of microtubules but appear to be largely independent of the distribution of actin. Thus, depolymerization of microtubules by microtubule-specific drugs stops particle movement, whereas depolymerization of actin filaments with cytochalasin D disrupts actin bundles but has little effect on saltatory particle movement (Menzel and Elsner-Menzel, 1989a).

Menzel (1987) has developed a microdissection technique that allows immunocytochemical procedures in *Caulerpa* without the necessity of digesting the cell wall enzymatically (Fig. 2.12). Using antibodies against actin and tubulin, three cytoplasmic levels can be identified in the cell area of the

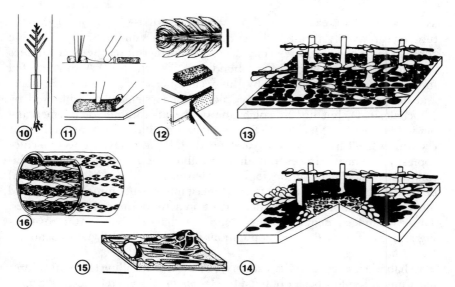

Figure 2.10. Schematic representation of a plantlet of *Bryopsis*. The rectangle denotes position of the artists views in Figs. 2.15 and 2.16 (Bar = 1 cm.) (After Menzel and Schliwa, 1986b.)

Figure 2.11. Scheme of the extrusion process of cytoplasmic cylinders from fixed cell segments of *Bryopsis*. (Bar for the lower scheme = 100 μm.) (After Menzel and Schliwa, 1986a.)

Figure 2.12. Apical portion of bilaterally organized *Caulerpa*, illustrated in face view (top). Rectangular cell area, to be removed from the uppermost, central portion of the thallus by microdissection, is marked (Bar = 4 mm.) Schematic representation of rectangular cell piece after dissection from the thallus in three-dimensional side view (center). Splitting of sandwichlike cell piece with the aid of a fixed razor blade, completely immersed in buffer (bottom). (From Menzel, 1987; with permission.)

Figure 2.13. Artists view of the cortex and a layer of subcortical microtubule bundles in *Caulerpa*. Immobile chloroplasts are indicated in black and mobile amyloplasts in white. The perpendicular columns represent cell wall trabeculae. Situation before UV-beam irradiation. (From Menzel and Elsner-Menzel, 1989b; with permission.)

Figure 2.14. Artists view of *Caulerpa*, similar to Fig. 2.13, but 4 minutes after continuous local UV irradiation. Chloroplasts have moved away from their original positions and gathered at the edge of the beam spot. Note that amyloplasts stop moving where the microtubule bundle is hit by the UV beam but continue movement in other areas. (From Menzel and Elsner-Menzel, 1989b; with permission.)

Figure 2.15. Artist's view of a segment of *Bryopsis*, taken out of the cortical cytoplasm as it appears after the chloroplasts have been removed by centrifugation, to demonstrate the spatial arrangement of spherical-shaped nuclei, rod-shaped mitochondria, and the channel system. The chloroplasts would be "swimming" on the surface of this layer. The widths of the cell wall and the cytoplasmic layer are not drawn to scale. (Bar = 10 μm.) (From Menzel and Schliwa, 1986a; with permission.)

Figure 2.16. Artist's view of the cell's architecture of *Bryopsis*, shown in Fig. 2.10. (Bar = 50 μm.) (From Menzel and Schliwa, 1986a; with permission.)

assimilatory blade, each with a distinct cytoskeletal organization. (1) Micro-tubules run axially through the cortex and form bundles with increasing complexity in the subcortex (Fig. 2.13). (2) Furthermore, the microtubules combine to form giant composite bundles in the center. (3) Actin, detected for the first time in *Caulerpa* by this technique, forms fine cortical fibers and filamentous foci (Fig. 2.14). Chloroplast movement was blocked by cytochalasin D, but not by colchicine or the microtubule-depolymerizing herbicide Cremart (Menzel and Elsner-Menzel, 1989b). The dynein inhibitor erythro–9-[3-(2-hydroxynonyl)] adenine (EHNA) also has no effect on chloroplast movement. However, both microtubule- and dynein-specific inhibitors block movement of amyloplasts. Menzel and Elsner-Menzel (1989b) demonstrated that, in contrast to amyloplast movement, immobilization and movement of chloroplasts are dependent on actin but not on microtubules. Thus, two independent mechanisms appear to have evolved for the positioning and motility of the two populations of plastids in the giant coenocyte *Caulerpa*.

Reliable evidence for the circadian clock as the controlling factor of chloroplast migration has been obtained in *Halimeda* (closely related to *Bryopsis* and *Caulerpa*), which becomes pale in the dark, remaining so for most of the night, but again is green enough by dawn to permit rapid photosynthesis as soon as light is available (Drew and Abel, 1990). This diurnal paling phenonemon affects the entire plant and has been observed in the field, in a laboratory seawater cascade system, and in laboratory experiments. Relocation of the chloroplasts to the segment surface well before dawn appears of some ecological importance in the field, where, at several meters depth, irradiance increases only slowly up to saturation level over a period of an hour or more after first light. Thus, the chloroplasts gather at periclinal walls that were, and presumably will be the next day, perpendicular to the light direction. In contrast to the situation here, chloroplast response to particularly bright light in many plants results in an aggregation in self-shading arrays at the anticlinal cell walls paralleling the light direction (see, e.g., Figs. 2.5 and 2.8; Haupt, 1982). Nultsch et al. (1981) concluded that the function of such reorientation was not the direct regulation of photosynthetic activity but the protection of the photosynthetic pigments from photo-damage.

Acetabularia

Similar to *Halimeda*, reliable evidence for a diurnal rhythm, regulated by the physiologic clock, has been detected in the giant unicellular green alga *Acetabularia mediterranea* (Koop et al., 1978). When the plantlets are still uninucleate (Fig. 2.17), a rhythmic greening and paling of the stalk, with a contrary effect in the rhizoid, can be observed. This again is due to displacement of the chloroplasts, and it may be added that even the nucleus, located in the rhizoid, performs small rhythmic displacements in the same direction as the chloroplasts. The oscillations persist nearly undamped in continuous light for at least seven diurnal cycles, with a period length of 25–27 hours (Fig. 2.21). There is no doubt, therefore, about the primary role of the physiologic clock in this movement, similar to other manifestations of the physiologic clock in this alga. A high-molecular-weight polypeptide of 230 kDa has been identified in *Acetabularia* that fulfills the requirements of an "essential clock protein" (Hartwig et al., 1985; 1986). The rate of synthesis of this protein oscillates under constant conditions, whereas the other proteins are synthesized with constant rates (de Groot and Schweiger, 1990).

In the vegetative state of *Acetabularia*, the cytoplasm is dominated by a cytoskeleton, composed of massive bundles of actin filaments embedded in a network of fine actin filaments (Figs. 2.18 and 2.19). The bundles form a parallel, axially oriented system running along the entire length of the cell, including the cap rays (Dazy et al., 1981; Menzel, 1986). Chloroplasts are transported along these bundles at a slow speed of 1–2 μm s^{-1}. Other organelles, including polyphosphate granules and a host of smaller, ill-defined vesicles, move at a high speed of 3–11 μm s^{-1} in parallel—much as traffic on the fast lane of a busy expressway (Fig. 2.20). The mechanisms underlying the two different modes of transport are not identified, although in the case of chloroplast transport the evidence points to an actin-based mechanism (Koop and Kiermayer, 1980b; Nagai and Fukui, 1981). The motor that drives the fast-speed traffic remains mysterious; the inhibition by colchicine and the microtubule-specific herbicide Amiprophosmethyl argue for an involvement of microtubules, but immunofluorescent data do not support this. Nagai and Fukui (1981) have shown that cytoplasmic streaming and chloroplast transport in the stalk of *Acetabularia* in the vegetative state are reversibly inhibited by cytochalasin B (CB) at 50 μg ml^{-1}, and irreversibly by N-ethylmaleimide (NEM) above concentrations of 0.25 mM. After spinning down the endoplasm to one end of the stalk by gentle centrifugation at about 500 × g for 3 min, numerous ectoplasmic striations remained in situ in the stalk cortex. The striations ran at unequal intervals in parallel with the longitudinal axis of the stalk (Figs. 2.18 and 2.19). The endoplasm streamed back only along these striations. By combining centrifugation and a double-chamber technique, the endoplasm and the cortex of the stalk were treated separately with CB and NEM. CB treatment of the cortex arrested streaming; when treatment was restricted to the endoplasm, streaming continued at a normal rate, NEM restricted to the cortex permitted normal streaming rates, and treatment restricted to the moving endoplasm inhibited streaming. These results suggest that microfilaments and possibly myosin play an active role in the streaming. Microfilaments must reside in the cortex, especially in the ectoplasmic striations, whereas the putative myosin must reside in the moving endoplasm.

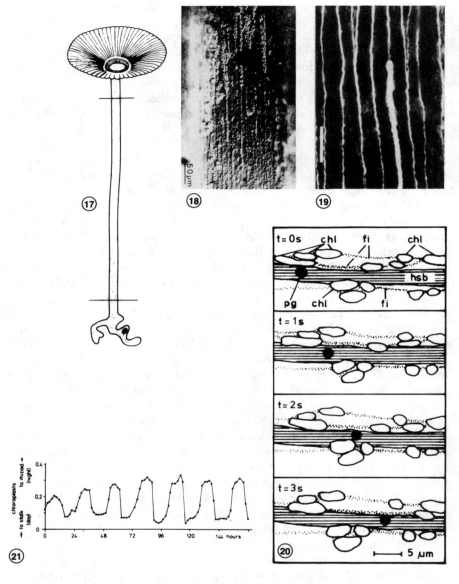

Figure 2.17. Scheme of an *Acetabularia* interphase cell. The horizontal bars indicate extension of the cylindrical stalk. (After Mohr and Schopfer, 1978.)

Figure 2.18. Part of stalk cytoplasm of *Acetabularia*, after centrifugation at about 500 × g for 3 minutes in the longitudinal direction. Lines of chloroplasts are seen to participate in the endoplasmic flow. (From Nagai and Fukui, 1981; with permission.)

Figure 2.19. Part of stalk cytoplasm of *Acetabularia*. Actin bundles visualized by indirect immunofluorescence using a monoclonal antibody against smooth muscle actin. Headed streaming band in center. (Bar = 50 μm.) (From Menzel and Elsner-Menzel, 1989c, with permission.)

Ulva

Like the multichloroplast systems referred to earlier, the single parietal chloroplast of the green alga *Ulva* performs diurnal movements in each cell of the two-layered thallus (Britz and Briggs, 1976). During the day the chloroplast faces the periclinal cell wall close to the thallus surface (i.e., perpendicular to the impinging light), and during the night the organelle is located at an anticlinal wall (see Fig. 2.8). Displacement to the night position starts long before the end-of-day signal, and vice versa. This phenomenon already hints at a circadian clock as the controlling factor. More convincingly, the movement persists for up to 10 days in continuous darkness or light (see Fig. 2.9) with a free-running period of 24–25 h (Britz et al., 1976). Inhibition of the movement by colchicine and its reversion by isomerization to lumicolchicine (Britz, 1976, 1979) provide evidence for microtubules being involved in chloroplast movement in *Ulva*. Interestingly, chlorophyll synthesis in *Ulva rigida* may be controlled by a phytochromelike photoreceptor that has been detected by the Western blot technique with monoclonal antibodies directed to phytochrome from etiolated maize and oat seedlings (Lopez-Figueroa and Niell, 1989; Lopez-Figueroa et al., 1989). The absorption difference spectrum after partial purification showed a "normal" absorption band ($\lambda_{max} = 670$ nm) for the P_r form but only a weak band ($\lambda_{max} = 705$ nm) for the "P_{fr}" form. So far, no phytochrome effect on circadian rhythm in *Ulva* has been shown.

In the majority of the case studies described so far, with *Ulva* possibly an intermediate situation, the chloroplast is moving as one organelle inter alia in "cytoplasmic streaming." This may be termed *chloroplast movement* in *sensu lato*.

The Chloroplast as Positioned
Independently from Other Organelles

Chloroplast movement, in the strictest sense, is characterized by sole chloroplast migration while the rest of the cytoplasm stays immobile. Light will

Figure 2.20. Artist's view of dual intracellular transport system in *Acetabularia*. Single-frame analysis of the movement of a polyphosphate granule in a "headed streaming band," which moves past chloroplast-containing cytoplasmic filaments. Abbreviations are the following: chl = chloroplast; fi = cytoplasmic filaments; hsb = headed streaming bands; pg = polyphosphate granule. (From Koop and Kiermayer, 1980a; with permission.)

Figure 2.21. Chloroplast movement in *Acetabularia* in continuous white light of irradiance as used for normal culture. Absorbance changes in the rhizoid region, given on the ordinate in relative units, indicate chloroplast movement in stalk and rhizoid direction, respectively, corresponding to day and night movement in the natural light–dark cycle. (After Koop et al., 1978.)

function here either to trigger chloroplast migration (the movement is performed in stereotypic fashion without respect to the light direction) or to govern chloroplast migration with respect to the light direction. The former photoresponse is realized in, for example, *Biddulphia, Eremosphaera,* and *Striatella*, whereas irradiance and direction of impinging light are responded to precisely by the single, ribbon-shaped chloroplast in the green algae *Mougeotia* and *Mesotaenium*.

Biddulphia, Eremosphaera, and *Striatella*

In the centric diatom *Biddulphia*, the unicellular green alga *Eremosphaera*, and in the pennate diatom *Striatella*, the nucleus is suspended by cytoplasmic strands in the center of the large vacuole. Along these strands, the chloroplasts migrate centripetally from the cortical cytoplasm to the nucleus. The chloroplasts are spread over the cell periphery in darkness and in low-irradiance light, without relationship to the light direction or to an absorption gradient. In high-irradiance light, on the contrary, the chloroplasts are triggered to aggregate around the centrally located nucleus (Haupt, 1983a; Weidinger and Ruppel, 1985).

Mougeotia and *Mesotaenium*

In sensory transduction processes and the clearly demonstrated photoresponse of these algae, at least three major components are needed to describe the reaction in sequential order: (1) a sensory pigment to perceive the light signal differentiated in wavelength and direction; (2) a sensory transducer converting the light signal into a chemical message; (3) an effector to respond to this message and to move the chloroplast according to the information perceived.

When the sensory pigment phytochrome is activated by low-irradiance white light or by red light in *Mougeotia* and *Mesotaenium*, the chloroplast will be oriented so as to achieve maximal light exposure. This behavior is termed *low-irradiance response* of the chloroplast to reach the "face-on" position (Fig. 2.22; Haupt, 1982). Additional sensory pigments, to cover the green and the blue spectral region of light, are likely (Figs. 2.24 and 2.25; Gabryś et al., 1984; Lechowski and Bialczyk, 1988; Kraml et al., 1988).

When blue and red light are given simultaneously (see Fig. 2.25)—for example, in high-irradiance white light—the chloroplast will be oriented in a second type of response to achieve minimal light exposure. This is called the high-irradiance response of the chloroplast to reach the "profile" or "edge-on" position (see Fig. 2.22; Schönbohm, 1980). Both types of photomodulatory response in *Mougeotia* and, with some quantitative difference, in *Mesotaenium*, start or stop almost without lag, depending on pigment activation, indicating a relatively direct transduction chain. Orientational movement of the chloroplast is performed within a minimum of 10 minutes.

In *Mougeotia* in particular, the in vivo properties of phytochrome have been studied in great detail (Fig. 2.23; Haupt, 1987), and the analysis has been extended to the reaction chain of the chloroplast response (Grolig and Wagner, 1988). Therefore, mainly *Mougeotia* will be discussed here, but chloroplast movement in *Mesotaenium* (and the fern protonema of *Adiantum*) will be included for comparison of dichroic phytochrome.

Dichroic Phytochrome. The photomorphogenic and photomodulatory pigment, phytochrome (Rüdiger, 1980), induces in plants diverse photoresponses, such as germination and flowering, but also chloroplast movement in *Mougeotia, Mesotaenium*, and *Adiantum* (Wada and Kadota, 1989). Recently, two different species of phytochrome—phytochrome I and phytochrome II (Furuya, 1989)—have been discovered with considerable differences between their amino acid sequence in higher plants, and a family of possibly five phytochrome genes has been identified in *Arabidopsis thaliana* (Sharrock and Quail, 1989). At present, no information is available on the action mechanism(s) of phytochrome and little information exists on the evolution of its gene(s) in higher and lower plants, including algae (Furuya, 1987; Dring, 1988; Wada and Kadota, 1989; Cordonnier et al., 1986; López-Figueroa et al., 1989). *Mougeotia* and *Mesotaenium* clearly show phytochrome-mediated responses, and *Mesotaenium* phytochrome has been purified and preliminarily characterized (Kidd and Lagarias, 1990).

By partial irradiation with a microbeam of a *Mougeotia* cell and evaluation of the chloroplast response (Fig. 2.26; Haupt, 1982), phytochrome has been shown here (1) to be located in the parietal cytoplasmic layer rather than in the chloroplast, (2) to act locally, independent of the chloroplast movement, without intracellular spreading. In addition to light *intensity*, the *Mougeotia* and *Mesotaenium* cells apparently sense light *direction*. Most important for unraveling the precise mechanism of this vectorial function in a small single cell is the discovery of an action dichroism of the *Mougeotia* phytochrome: In linearly polarized light, a chloroplast response is shown only in cells for which the electrical vector of the light (the vibration plane) is perpendicular to the long axis of the cylindrical cell (Haupt, 1982). This finding led to the conclusion that the phytochrome molecules serving as pigment for the response are oriented in a dichroic pattern (Haupt, 1982). This pattern has been analyzed further by proper combination of irradiation with polarzied light, especially to small areas (see Fig. 2.26). Accordingly, the transition moments of the red-absorbing phytochrome (P_r) molecules are oriented parallel to the cell surface in a helical pattern at the periphery of the cell cylinder (Fig. 2.27).

When the physiologically inactive red-absorbing form of phytochrome (P_r)

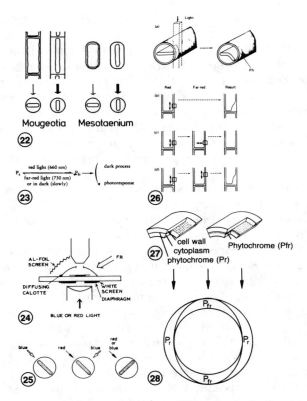

Figure 2.22. Chloroplast orientations in *Mougeotia* and *Mesotaenium*. Surface views and cross sections through a cell in unilateral white light of low irradiance (light arrow) and of high irradiance (heavy arrow). In the surface views, the light direction is normal to the plane of the paper. (After Haupt, 1987.)

Figure 2.23. Scheme on photoreactions of phytochrome and dark reversion. An absorbed photon of red light (λ max = 660 nm), converts to P_r within milliseconds into P_{fr}. Similarly, when P_{fr} absorbs a photon of far-red light (λ max = 730 nm), it is reverted to P_r. Such a weavelength-dependent shift in pigment feature is denoted "photochromic" (Hartmann, 1983). In simple terms, phytochrome is generally considered to function as a reversible biological switch, with P_r as the inactive form and P_{fr} as the active form. P_{fr}-induced photoresponses are canceled rapidly by far-red light or slowly in the dark (dark reversion).

Figure 2.24. Example of experimental set-up for the irradiation of trichomes of *Mougeotia*. Orienting blue or red light is given through the condenser (arrow) of the microscope, far-red light from the side, in this case reflected by an aluminum (AL) foil screen, and scattered by a celluloid calotte. The algae are placed on the glass slide below the cover slip. (From Gabryś et al., 1984; with permission.)

Figure 2.25. Single- and double-irradiation experiments of a trichome of *Mougeotia* in cross section, demonstrating chloroplast orientations in response to high-irradiance vectorial blue light (left), high-irradiance scalar blue light together with vectorial red light (center), and vectorial red or vectorial blue light of low irradiance (right).

Figure 2.26. Local irradiation of a *Mougeotia* cell with a microbeam of light (left) and response of the chloroplast (right). (a) General view; (b–d) the individual irradiation protocols. Two-headed arrows indicate the electrical vector of the linearly polarized light in the microbeam. Induction of chloroplast response by parallel vibrating red light indicates surface-parallel P_r; its

is photoconverted into the active far-red-absorbing form (P_{fr}), the transition moment of the chromophore changes: from surface-parallel orientation for P_r to surface-normal orientation for P_{fr} (see Fig. 2.27). In consequence, alignment of the transition moment of phytochrome with the direction of impinging light will result in a tetrapolar gradient of P_{fr}, with P_{fr} concentration highest at front and rear, where light hits the cell cylinder perpendicularly, but lowest at the two flanks (P_r) (Fig. 2.28). As a unique phenonemon in *Mougeotia*, the tetrapolar P_{fr}/P_r-gradient persists in darkness over several hours. Thus, there must be some specific structure–mechanism to prevent P_r versus P_{fr} from lateral diffusion over that long period (Fig. 2.31). A similarly fixed arrangement of P_r versus P_{fr} has not been reported for *Mesotaenium* or for fern protonemata of *Adiantum*, where absorption dichroism of P_r versus P_{fr} was shown as well (Haupt, 1987; Yatsuhashi et al., 1987).

Absorption change, caused by differences in the dipole moments of P_r and P_{fr}, was photometrically detected in phytochrome purified from higher plants and immobilized on agarose beads (Sundqvist and Björn, 1983a,b). From light-induced changes in linear dichroism at 730 nm and 660 nm, a rotational angle of about 31° (or in complement, 149°) was calculated (Ekelund et al., 1985). Although this value is small compared with the correlative physiological data (a difference of 90°), it is more or less compatible with the angle deduced from calculations based on physiological data for *Mougeotia* and for *Adiantum* (Björn, 1984; Sugimoto et al., 1987). The rate of the reorientation of phytochrome is sufficiently slow that the change can be ascribed to the conformational change of the protein moiety (but not of the chromophore), or to a change in the interaction between phytochrome and the receptor connecting phytochrome and the membrane, if it exists (Kadota et al., 1986).

As a consequence, *Mougeotia* makes elegant use of the two properties of phytochrome—that is, photochromism *and* change in absorption dichroism, with the photochromic equilibrium $P_r \rightleftharpoons P_{fr}$ (see Fig. 2.23) also becoming dependent on the azimuth of the impinging light, even under saturating light conditions. The ordered and obviously stable positioning of phytochrome appears indicative of the photoreceptor attached to the plasma membrane

---

reversion by perpendicularly vibrating far-red light indicates surface-normal P_{fr}. (From Haupt, 1983b; with permission.)

Figure 2.27. Artists view of part of the cell wall and the cytoplasm of *Mougeotia* with phytochrome molecules at the cytoplasmic membrane (dashes). Orientation of transition moment of phytochrome parallel (P_r) or normal (P_{fr}) to the cell surface. (From Haupt, 1983b; with permission.)

Figure 2.28. Schematic cross section through a *Mougeotia* cell, with phytochrome locally being mainly in the P_{fr} and the P_r form, respectively, as a result of irradiation by proper unidirectional (= vectorial) light (arrows). Phytochrome pattern (tetrapolar P_{fr}/P_r-gradient) is indicated by the ellipsoid, but other shapes are possible. (After Haupt, 1983b.)

and/or possibly to cytoskeletal structures. Microtubules, however, found in abundance just beneath the plasmalemma in the form of hoops oriented perpendicularly to the cell length axis of *Mougeotia* (Galway and Hardham, 1986), seem to be of minor importance, as microtubule assembly inhibitors such as colchicine, nocodazole, or vinblastine left the chloroplast response undiminished but doubled the speed of movement (Serlin and Ferrell, 1989). The conclusion of plasmalemma-associated phytochrome was drawn also from growth-related responses [i.e., polarotropism in *Mougeotia* (Neuscheler-Wirth, 1970) or in fern chloronemata (Etzold, 1965) and fern protonemata (Kadota et al., 1984)].

From physiological evidence, the action of phytochrome in light-induced chloroplast movement in *Mougeotia* and *Mesotaenium* appears well established (Haupt, 1987). A soluble fraction of *Mesotaenium* phytochrome was isolated and purified to apparent homogeneity (Kidd and Lagarias, 1990). Immunoblot analyses using a cross-reactive pea phytochrome monoclonal antibody revealed that (1) the 120-kDa band represents the full-length polypeptide; (2) phytochrome is predominantly (80 percent) localized in the algal cytoplasm; (3) there are 150,000–250,000 phytochrome molecules per cell. Interestingly, the yield of phytochrome molecules per cell increased fourfold when the cell material was kept in darkness for five days prior to phytochrome isolation (Kidd and Lagarias, 1990), reminiscent of similar effects reported earlier from tissue of etiolated higher plants (Quail, 1984). With respect to the well-documented action dichroism in chloroplast response (see earlier), the residual fraction of nonsoluble phytochrome in *Mesotaenium* may turn out to be physiologically more relevant.

Mechanics of the Movement. After changing the phytochrome gradient by changing the direction of incident light, the chloroplast edges, which are in contact with the cortical cytoplasm, move randomly in either P_r direction. A transient chloroplast bending may result; if so, it proceeds until one edge stops and turns back all the way. The reoriented chloroplast finally flattens.

As time-lapse fixation during movement and subsequent thin sectioning have revealed (Fig. 2.29), the chloroplast edges (like the entire chloroplast) are soft and often lobed during movement. The first evidence that the motor apparatus was actin–myosin, came from the use of drugs, including cytochalasin B (CB). CB turned out to be a rather specific and quickly reversible probe: CB was found to inhibit chloroplast movement, not however, light perception (Wagner and Klein, 1981). This result was the precondition to test at physiological temperatures of 20°–22°C, if, and for how long, the intracellular P_{fr} gradient stores the information on light direction. The results showed that the information disappears in darkness according to a first-order reaction with a half-life of about 90 minutes. This half-life is in the same time range as typical for the dark reversion of P_{fr} in vivo (e.g., Furuya,

1989). Moreover, even after 30 minutes with CB (Fig. 2.31), induction of chloroplast movement can be completely reversed by a short far-red irradiation. The data are consistent with similar results after an intermittent period of inhibited chloroplast movement by cold temperature (Haupt, 1982). Hence, we conclude in *Mougeotia* that the P_{fr} gradient itself functions as memory and stores the information on light direction, stabilized by some unknown structure–mechanism to prevent P_r versus P_{fr} from lateral diffusion. The memory in darkness, and thus fixation of the P_{fr} gradient, is less stable in *Mesotaenium* and the fern protonemata of *Adiantum* (see earlier). Marchant (1976) succeeded in decorating actin filaments by heavy meromyosin in a cell homogenate, as did Klein and co-workers (1980) in spread protoplasts of *Mougeotia*. 5–10-nm-thick filaments could be demonstrated in the cortical cytoplasm running in between the chloroplast edge and the plasmalemma (Wagner and Klein, 1978). Positioned as they are (Figs. 2.30 and 2.33), these filaments may well be able to move the chloroplast edge, which is then pulled into lobes, because of its softness (Fig. 2.29). Even so, the most intriguing aspects of the light-oriented and sensor pigment–mediated chloroplast response remain to be clarified: the mechanisms of initiation and of coordination of the movement.

Calcium appears involved here but direct proof is missing. Namely, the six paradigms of Williamson (1981) that identify calcium as a secondary messenger are not yet valid for *Mougeotia* or *Mesotaenium*. Rather, several lines of evidence exclude a coordinative function of calcium in chloroplast movement.

Enhanced calcium flux through the *Mougeotia* plasmalemma in continuous white light and a significant decrease in intermittent red light, related to photosynthesis, have been reported (Wagner and Bellini, 1976). Red–far-red reversible uptake of $^{45}Ca^{2+}$, measured by microautoradiography in single *Mougeotia* filaments, appears to be P_{fr}-dependent (Fig. 2.36; Dreyer and Weisenseel, 1979), but may be restricted to the calcium-rich pectic layer of the cell wall (Grolig, 1986). Roux (1984) reported a red-light-induced Ca^{2+} efflux from dark-adapted *Mougeotia* cells as determined by the low-affinity Ca^{2+}-indicator murexide. To possibly mimic active phytochrome by calcium in *Mougeotia* (see Williamson, 1981), Serlin and Roux (1984) successfully used the Ca^{2+} ionophore A 23187 in part of the experiments. However, P_{fr} mimicry by plasmalemma flux of calcium was not possible under all conditions tested (Serlin and Roux, 1984). When the ionophore was applied unilaterally, there was no chloroplast response, in contrast to the result when unilateral red light was given (Fig. 2.39). Additionally, patch clamp data from *Mougeotia* protoplasts clearly separate cation channel activity from chloroplast movement (Lew et al., 1990b). Thus, the primary action of phytochrome in chloroplast movement in *Mougeotia* is not mimicked by plasmalemma calcium flux.

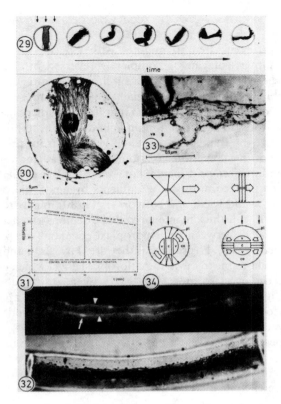

Figure 2.29. Representative drafts from electron micrographs on time-lapse sequence (horizontal arrow) of chloroplast movement in *Mougeotia* after proper light induction (vertical arrows). The cross sections at the left and right ends (namely, at rest *before* and *after* reorientation) show the chloroplast as a flattened, almost rectangular organelle to transverse the cylindrical cell. During movement (central drafts), the chloroplast edges appear bent and often lobed. (After Wagner and Klein, 1981.)

Figure 2.30. Electron micrograph of a cross section of a *Mougeotia* cell showing some details of the chloroplast (chl), including a pyrenoid (py); the adjacent cytoplasm (cy) contains dictyosomes (di) and endoplasmic reticulum (ER). Further details are as follows: va = divided vacuole; co = cortical cytoplasm; cw = cell wall. (After Wagner and Klein, 1981.)

Figure 2.31. Reversion of red-light-induced chloroplast movement in *Mougeotia* by means of far-red light, applied immediately before washing out of the actin-specific inhibitor cytochalasin B (CB; concentration = 7.5 μg ml^{-1}). Ordinate: Percentage of responding chloroplasts after removal of CB. Abscissa: After the red-light induction, time in darkness at which CB has been removed by washing out. (From Wagner and Klein, 1981; with permission.)

Figure 2.32. Localization of F-actin in *Mougeotia* as visualized after binding of rhodamine-phalloidin (RLP). Fluorescent (top) versus transmittent view (bottom) of identical sections of a *Mougeotia* cell, connected end-on within the trichome, in the light microscope. Mostly a pair of fringes of diffuse RLP-fluorescence adjacent to each longitudinal chloroplast edge is seen (arrow heads), together with a number of fluorescent patches near the chloroplast edge (arrow). (from Grolig et al., 1990; with permission.)

Figure 2.33. Electron micrograph of a chloroplast edge from *Mougeotia* in cross section. Filaments (fi) can be identified running in between the chloroplast edge (chl) and the plas-

For spatial information to become adequate for the actomyosin motor apparatus in *Mougeotia*, and possibly in *Mesotaenium*, the primary signal stored in the photoreceptor memory of active phytochrome or blue-light pigment must be transformed into the corresponding pattern of the structurally fixed component of the motor apparatus. A likely candidate appears to be the plasmalemma anchorage sites to actin filaments (Grolig and Wagner, 1988).

Alignment of the actin filaments at sites where chloroplast edge and cortical cytoplasm merge (Wagner and Klein, 1981) shows that parts of the actomyosin-generated force are oriented in a radial direction within the cylindrical cell. Chloroplast orientation, however, results from force in a tangential direction. Hence, chloroplast movement demands successive lateral shifts of the actin filaments that presumably proceed in a statistical way, biased by sensor pigment–modulated plasmalemma anchorage sites to actin. Consequently, the velocity of *Mougeotia* and *Mesotaenium* chloroplast reorientational movement does not result merely from the velocity of actin–myosin interaction, as is the case in translational movements (e.g., cytoplasmic streaming). Rather, rotational movement may be dominated by biased, lateral movement of actin filaments along an increasing gradient of plasmalemma anchorage sites. Actually, the velocity of reorientational movement of the *Mougeotia* chloroplast (about 10^{-8} m s^{-1}) turns out to be two to three orders of magnitude slower than the translational actin-based intracellular movements (Williamson, 1975). Furthermore, as shown by electron microscopy or after rhodamine phalloidin staining of formaldehyde-fixed cells (Fig. 2.32), a special F-actin organization of seemingly single filaments is evident for walled *Mougeotia*, in contrast to actin bundles in identically treated *Mougeotia* protoplasts (Grolig et al., 1990). It may be speculated that actin filament *bundles* would not allow enough freedom for the single actin filament to undergo the biased, lateral stepwise movement as implied in the model.

In the hypothetical mechanism shown here (Fig. 2.34), actomyosin interconnecting the chloroplast envelope and the plasmalemma provides force for holding the chloroplast in position and enables reorientation of the

malemma (pl) in the flap of cortical cytoplasm. Details are abbreviated as in Fig. 2.30, except the tonoplast (to) and the plasmalemma (pl). (From Wagner and Klein, 1981; with permission.)

Figure 2.34. Top: Linear model of movement based on a gradient of anchorage sites. Movement is biased upon statistical migration of shearing filaments reversibly binding to the anchorage sites. Open arrows indicate resulting direction of movement. Bottom: Adaptation of the linear model to geometric conditions as in the cylindrical *Mougeotia* cell. Note the random reorientation of the chloroplast edge at the beginning of the movement (hatched lines), resulting occasionally in a transient bending of the cloroplast (cf. Fig. 2.29). Mostly, the chloroplast edge pointing to the light source takes the lead. Direction of proper unidirectional light is also shown (small arrows). Cell structural details are as follows: c = chloroplasts; co = cortical cytoplasm; pl = plasmalemma; v = vacuole. (From Grolig and Wagner, 1988; with permission.)

chloroplast edges according to plasmalemma-stored, spatial information. Thus, actin, together with plasmalemma anchorage sites, could serve both functions and might work without an additional trigger (e.g., Ca^{2+}). However, a second degree of freedom seems to be needed to encompass additional phenomena. After red or white light induction of the low-irradiance response in edge-on preoriented and then dark-adapted cells, a 5-minute delayed increase of chloroplast adhesion to cortical cytoplasm (Schönbohm et al., 1990) is observed in centrifuged cells (Fig. 2.37), followed by a decrease of adhesion after 20 minutes to the dark level as seen before induction. Similar results were obtained from *Mougeotia* cells after similar light treatment, but with the chloroplast pre-oriented in the face-on position (Schönbohm, 1973). Thus, there must be an additional factor involved to impede (e.g., in darkness) the restraining function of actomyosin. Such a factor may well be the activated state of actomyosin, depending, for example, on cytoplasmic pCa. Thus, for the actomyosin motor apparatus, two steps of regulation by the sensor pigment(s) may be assumed (i.e., gradient formation of plasmalemma anchorage sites to actin filaments followed by initiation of actomyosin interaction). Consequently, the state of preparedness of the individual algal cell or the algal trichome determines whether the actomyosin interaction was initiated long before gradient formation of plasmalemma anchorage sites to actin filaments.

Regulation—Steering of the Movement. *Mougeotia* cells contain and release phenolic compounds to the external medium (Schönbohm and Schönbohm, 1984). These probably act as chemical defense against herbivores or microbial attacks and as growth-inhibiting substances (Menzel, 1988). The phenolics probably originate from cytosomes (Schönbohm and Schönbohm, 1984), which seem to be identical with the calcium-binding vesicles reported by Wagner and Rossbacher (1980) and analyzed by Grolig and Wagner (1987, 1989; see also later). Crude separation of the phenolic compounds released to the external medium revealed fractions with strong inhibitory effect on the high-irradiance response but left the low-irradiance response almost unimpaired (Schönbohm and Schönbohm, 1984). This finding adds evidence to the idea that the chloroplast high-irradiance response in *Mougeotia*, and possibly in *Mesotaenium*, results from a more complex and therefore more readily disturbed sensory transduction chain as compared to the low-irradiance response.

Schönbohm and co-workers (1990) did an extensive study of the effect of Ca^{2+} entry blockers on light-induced chloroplast movements and on chloroplast adhesion in *Mougeotia* (Fig. 2.37). The organic inhibitors diltiazem and nifedipine, and the inorganic inhibitors ruthenium-red, La^{3+} (see also Wagner and Grolig, 1985) and Co^{2+}, did not affect the low- and high-irradiance movements of the chloroplast. Only at toxic concentrations, or after long-term incubations of 4–7 days (which resulted in unspecific side effects), was the chloroplast movement slightly inhibited. These data contradict the hypothesis of a P_{fr}-controlled plasmalemma-bound Ca^{2+} influx as an essential link in the signal transduction chain of light-oriented chloroplast movements in *Mougeotia* (Schönbohm et al., 1990). This conclusion falls in line with other evidence: Serlin and co-workers (Lew et al., 1990a,b) have used the patch clamp technique to determine whether phytochrome regulates ion channel activity in the plasmalemma of *Mougeotia* protoplasts. Consistent with the pharmocologic results of Schönbohm et al. (1990) and of Wagner and Grolig (1985), the patch clamp data of Serlin and co-workers (Lew et al., 1990b) indicate that "it is unlikely that channel activation is part of the mechanism leading to chloroplast rotation . . . in *Mougeotia*."

Calcium-binding vesicles have been discovered in *Mougeotia* (Wagner and Rossbacher, 1980; Grolig and Wagner, 1987, 1989). They are abundant at the chloroplast edges, where they accumulate even more during reorientational movement (Wagner and Klein, 1981). Since these vesicles contain phenolics, they specifically adsorb vital dyes [(e.g., neutral red (NR) and rhodamine B)] in vivo (Grolig and Wagner, 1987). Vital staining has permitted isolation of the otherwise disintegrating vesicles (Figs. 2.35 and 2.38) and has enabled their characterization in vitro (Grolig and Wagner, 1989). They were found to possess a protonated group with $pK_8 = 9.9$, typifying phenolic hydroxyl groups; upon titration, both phenolic compound(s) and vital dye were concomitantly released from the vesicle matrix. A shift in peak absorption, from 450 to 540 nm, of neutral red upon adsorption to the vesicles indicated that the neutral form of neutral red was bound to the vesicular matrix as an intermediate form, stabilized via intermolecular hydrogen bonds to the phenolic compound(s). Analysis of Langmuir adsorption isotherms indicated that there were two binding sites each for both neutral red and rhodamine B. The isolated vesicles were devoid of calcium, probably because vesicular calcium, bound to the vesicle matrix, was displaced upon dye binding (Grolig and Wagner, 1989).

In intact *Mougeotia* cells, increasing amounts of vesicularly bound neutral red or rhodamine B lead to increasing inhibition of the chloroplast movement. Locally restricted staining of a *Mougeotia* cell caused locally restricted inhibition of the chloroplast movement, which could be overcome by the cell with time (Fig. 2.41). In fact, within an intermission of 40 minutes in white light, the inhibited part of the chloroplast recovered full capability for reorientational movement, in spite of the persisting staining of the vesicles (Russ et al., 1988). The date on NR-inhibited chloroplast movement (Grolig and Wagner, 1989) and on NR-induced vesicular calcium release lead to the following interpretation: (1) The dye, upon vesicular binding, probably causes effusion of vesicular calcium, which overrides the cytoplasmic calcium homeostasis (Russ et al., 1988); (2) despite the persistent staining, the cell

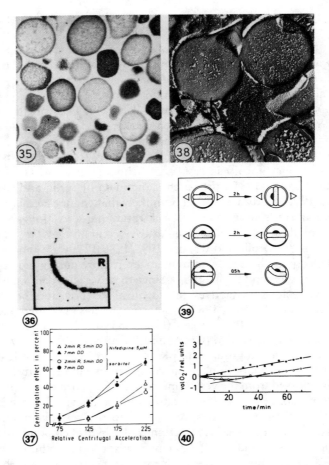

Figure 2.35. Electron micrograph of calcium-binding vesicles from *Mougeotia*, isolated after stabilization in vivo by staining with rhodamine B. The osmiophilic outer portion of the matrix and a nonosmiophilic core can be distinguished. Magnification = 10,000 ×. (F. Grolig, unpublished.)

Figure 2.36. Representative autoradiograph from partially irradiated trichome of *Mougeotia* (see rectangle), illustrating the effect of a 3-second irradiation with red light (R) on accumulation of $^{45}Ca^{2+}$. (From Dreyer and Weisenseel, 1979; with permission.)

Figure 2.37. The effect of nifedipine on chloroplast adhesion in dark-pretreated (DD, filled symbols) versus red-irradiated *Mougeotia* trichomes (R, open symbols). The trichomes were exposed to 5 μM nifedipine for 48 hours. Sorbitol = control. The bars indicate standard error of the mean. (From Schönbohm et al., 1990; with permission.)

Figure 2.38. Electron micrograph of freeze-etched calcium-binding vesicles from *Mougeotia*, isolated after stabilization in vivo by staining with neutral red. Compare the structural details of the vesicle matrix with those discernible in Fig. 2.32. Magnification = 16,600 ×. (F. Grolig, unpublished.)

Figure 2.39. Scheme of cross section of a *Mougeotia* cell. Left-hand side: (top) bilateral positioning, (center) unilateral positioning of the calcium ionophore A 23187, and (bottom) unilateral positioning of a beam of red light. Right-hand side: chloroplast reorientation after the indicated

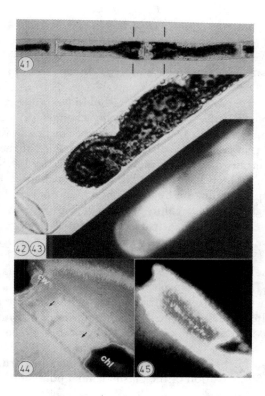

dark periods. The unilateral red light (bottom) results in local reorientation (cf. Fig. 2.36) in contrast to unilateral application of the calcium ionophore (center). [After Serlin and Roux, 1984 (top and center) and Haupt, 1983b (bottom), respectively.]

Figure 2.40. Rate of photosynthetic O_2 production in *Mougeotia* at low-irradiance white light as a function of chloroplast position. Closed circles indicate cells with chloroplasts in "face on" position during the whole of the experiment. Open circles indicate cells, at the start of the experiment, with chloroplasts in "edge-on" position but turning to "face-on" position during the experiment. O_2 production passes the compensation point when the chloroplasts turn from "edge-on" to "face-on" position. (After Zurzycki, 1955.)

Figure 2.41. Microview of two *Mougeotia* cells within a trichome. Upon proper light induction, the chloroplasts performed reorientational response into "edge-on" position at the left- and the right-hand side of the cells, respectively, and the central part remained in "face-on" position because of local application of neutral red at both flanks (see marked section). Note also the gathering of vesicles where neutral red had been applied. Within an intermission of 40 minutes in white light (not shown), the inhibited part of the chloroplast recovered full capability for reorientational movement, despite the persisting aggregation of the stained vesicles. (U. Russ, unpublished.)

Figures 2.42 and 2.43. Brightfield and fluorescence micrograph of an indo-1-loaded plasmolyzed *Mougeotia* cell as part of the trichome. The fluorescence originates from the protoplast of the cell, partly quenched by the chloroplast, but not from the cell wall. Chlorophyll fluorescence was damped with a colorglas. (U. Russ, unpublished.)

Figures 2.44 and 2.45. Pseudocolor micrographs of a *Mougeotia* cell within the trichome after loading of indo-1. The chloroplast has been dislocated at one end of the cell by centrifugation (Fig. 44), leaving one cytoplasmic strand (arrows) running from the chloroplast (chl) to the crosswall (cw). The highest indo-fluorescence (pink, Fig. 45) originates from this strand. Chlorophyll fluorescence was suppressed with a glass filter. (U. Russ, unpublished.)

regains cytoplasmic calcium homeostasis by cellular export of this regulatory ion.

The vesicles, acting as a calcium buffer, should guarantee calcium homeostasis in *Mougeotia* cells—readily balancing changes in cytoplasmic pCa (Grolig and Wagner, 1989). In preliminary experiments, however, red–far-red–dependent changes in the calcium buffer capacity of the vesicles have been found by means of chlorotetracycline fluorescence in vivo (Wagner et al., 1987). Also, an increase in cytoplasmic free calcium in *Mougeotia* has been detected upon irradiation by blue light of 365 nm in a dose as used for induction of high-irradiance chloroplast movement. For the latter experiments, the fluorescent calcium-sensitive dye indo–1 was loaded for 2 hours (Figs. 2.42 and 2.43), with most of the indo-1-fluorescence originating from the cytoplasm (Figs. 2.44 and 2.45; Russ et al., 1991). This cytoplasmic calcium increase turned out to be independent of the external calcium concentration (EGTA-buffered media of pCa 8 versus pCa 3), and probably reflects calcium release from internal stores. It was suggested that blue-or red-light-mediated release of calcium from internal stores may result in at least two effects: (1) depolymerization of microtubules (Russ et al., 1991), which decreases cytoplasmic viscosity (Weisenseel 1968; Schönbohm and Hellwig, 1979), much like microtubule-depolymerizing drugs (Serlin and Ferrell, 1989); (2) activation of the plasmalemma potassium channels (Lew et al., 1990b).

Either calcium by itself or calcium-activated calmodulin, isolated from *Mougeotia* and *Mesotaenium* (Wagner et al., 1984; Jacobshagen et al., 1986), should be able to pass the Ca^{2+} signal onto cytoskeletal regulatory proteins [e.g., by activation of protein kinases (Wagner et al., 1987)].

By using a synthetic peptide, KM–14, Roberts (1989) has detected and partially purified a protein kinase from *Mougeotia*. The peptide contains a sequence of the regulatory light chain of smooth muscle myosin which is phosphorylated by calcium-calmodulin-dependent myosin light-chain kinase (MLCK). The *Mougeotia* kinase was stimulated 40-fold by calcium, with half-maximal stimulation occurring at 1.5 μM. The calmodulin-depleted enzyme was fully active and calcium dependent and was not stimulated further by exogenous calmodulin or by the calcium effectors phosphatidylserine and diacylglycerol. The enzyme phosphorylated intact chicken gizzard myosin light chain, consistent with rabbit myosin light chain (Wagner et al., 1987) as well as the KM–14 substrate. KM–13, a peptide analog of KM–14 with a deletion of a glutamine at position 5, was a poor substrate with a V_{max}/K_m ratio 200-fold lower than that of KM–14. Calcium-dependent, but not calcium-calmodulin-dependent, KM–14 kinase activity was also detected in *Mesotaenium*. On the other hand, inhibition of the reorientational movement of the *Mougeotia* chloroplast (after induction of the low-irradiance response by blue or red light) by the calmodulin antagonists trifluoperazin and the W-compounds (Wagner et al., 1984; Serling and Roux, 1984) suggest the involvement of calmodulin (Wagner et al., 1987). Despite the specificity of the drugs used, however, the effect of calmodulin on chloroplast movement in *Mougeotia* and possibly *Mesotaenium* may be indirect, since the regulatory functions of calmodulin in the cell are multiple (Dieter, 1984).

Conclusion

Chloroplast motility in the various algal systems described here utilizes the two basic motility systems in plants, with microtubules and microfilaments as the vector-orienting components along which the associated force-generating molecules move. The two systems act in parallel in *Caulerpa* to position the two populations of plastids (i.e., chloroplasts and amyloplasts). In *Acetabularia* and *Bryopsis* the two systems may act together in the transport of organelles. Otherwise, one of the two motility systems dominates with consistent functional correlations: The microtubule-based system is at work in the circadian movements entrained but not directed by light, whereas the orientational movements with respect to the light direction always rely on the actin-based motility system.

Chloroplast migration, as far as is known, is found only in nonmotile organisms which keep their cellular position fairly constant within the environment. The impact of light on chloroplast positioning, possibly relevant for the photosynthetic yield (see later), depends on the various features of the light sensed by the cells: (1) light *quantity* (and *quality*) may entrain the endogenous clock or switch between two structurally predetermined chloroplast positions; (2) in addition, light *direction* is sensed by dichroic sensory pigments possibly associated with the plasma membrane, and light-oriented chloroplast positioning will result.

The ecological significance of the chloroplast migration is not easily envisaged in some of the systems. For example, for the rhthmic chloroplast movements in *Ulva*, unrelated to the light direction, there is no apparent use, although protection of these photosynthetic organelles against photodamage may be a major function. On the other hand, nightly withdrawal of the chloroplasts into the substrate-hidden parts of the thallus, as observed in *Halimeda* and *Acetabularia*, is apparently used to protect the light-harvesting "flock" of organelles against predators.

Chloroplast movements have evolved despite the energy that is consumed for force generation during movement and positioning. However, one must be aware that this amount of energy is only a small fraction of the total energy consumption, which is mainly used up in the anabolic pathways of these biosynthetically active cells (Berg, 1983). With respect to this small

energy expenditure, the cellular function of light-governed chloroplast positioning in some systems may indeed be a tolerated one, of minor positive
influence on photosynthetic yield but also without significant disadvantage
(waste of energy) for the cell. In other systems, light-triggered chloroplast
rearrangement, irrespective of light direction, may prevent photodamage of
the then often clustered and self-shading chloroplasts or of a hidden nucleus.

The most sophisticated chloroplast movements (e.g., seen in *Mougeotia*
and *Mesotaenium*) are mediated by several sensory pigments to cover the full
spectrum of light. The photochromic pigment phytochrome, predominant in
governing chloroplast reorientation in *Mougeotia* and *Mesotaenium*, provides the cell with proper information on the quality and direction of light
available for photosynthesis. Directionality is imprinted here in a certain
pattern of activated photoreceptor molecules, easily switched on by light
stimulus in ultrafast photoreactions, in concert with slow-responding cytoskeletal components. With high sensitivity and precision, phytochrome
detects, for example, spectral changes of light attenuation by chlorophyll,
resulting in the "green shade." This green shade is the normal situation in
the algal aggregates in the open environment, with criss-crossed filaments
and locally (even within a single cell) highly variable light conditions. In
contrast to the other systems described, the ecological significance of the
chloroplast movement under low-irradiance conditions has been tested successfully in *Mougeotia*: The net rate of assimilation is negative if the edge
of the ribbon-shaped *Mougeotia* chloroplast points to the incident low-
irradiance light but increases to reach positive values when the chloroplast
turns into the face-on position (Fig. 2.40; Zurzycki, 1955).

ACKNOWLEDGMENTS

We greatly acknowledge skillful photographic assistance by A. Weisert.
Work in the authors' laboratory has been supported financially by the
Deutsche Forschungsgemeinschaft.

References

Berg, H. C. 1983. *Random Walks in biology.* Princeton University Press, Princeton.
Björn, L. O. 1984. Light-induced linear dichroism in photoreversibly photochromic
    sensor pigments. V. Reinterpretation of the experiments on *in vivo* action dichro-
    ism of phytochrome. *J. Plant Physiol.* 127:187–191.
Blatt, M. R. 1983. The action spectrum for chloroplast movements and evidence
    for blue-light-photoreceptor cycling in the alga *Vaucheria. Planta* 159:267–
    276.

Blatt, M. R., and Briggs, W. R. 1980. Blue-light-induced cortical fiber reticulation concomitant with chloroplast aggregation in the alga *Vaucheria sessilis*. *Planta* 147:355–362.

Blatt, M. R., Weisenseel, M. H., and Haupt, W. 1981. A light-dependent current associated with chloroplast aggregation in the alga *Vaucheria sessilis*. *Planta* 152:513–526.

Blatt, M. R., Wessells, N. K., and Briggs, W. R. 1980. Actin and cortical fiber reticulation in the siphonaceous alga *Vaucheria sessilis*. *Planta* 147:363–275.

Britz, S. J. 1976. Studies on a circadian rhythm of chloroplast movement in *Ulva*. Ph.D. thesis, Harvard University, Cambridge, MA.

Britz, S. J. 1979. Chloroplast and nuclear migration. In A. Pirson and M. Zimmerman (eds.), *Encyclopedia of Plant Physiology: New Series*, Vol. 7. Springer, Berlin, pp. 170–205.

Britz, S. J., and Briggs, W. R. 1976. Circadian rhythms of chloroplast orientation and photosynthetic capacity in *Ulva*. *Plant Physiol.* 58:22–27.

Britz, S. J., Pfau, J., Nultsch, W., and Briggs, W. R. 1976. Automatic monitoring of a circadian rhythm of change in light transmittance in *Ulva*. *Plant Physiol.* 58:17–21.

Cordonnier, M.-L., Greppin, H., and Pratt, L. H. 1986. Identification of a highly conserved domain on phytochrome from angiosperms to algae. *Plant Physiol.* 80:982–987.

Dazy, A.-C., Hoursiangou-Neubrun, D., and Sauron, M. E. 1981. Evidence for actin in the marine alga *Acetabularia mediterranea*. *Biol. Cell.* 41:235–238.

de Groot, E. J., and Schweiger, M. 1990. Expansion of the "coupled translation-membrane model" of circadian rhythm prokaryotes. *Bot. Acta* 103:5–6.

Delbrück, M., and Shropshire, W., Jr. 1960. Action and transmission spectra of *Phycomyces*. *Plant Physiol.* 35:194–204.

Dieter, P. 1984. Calmodulin and calmodulin-mediated processes in plants. *Plant Cell Environ.* 7:6:371–380.

Drew, E. A., and Abel, K. M. 1990. Studies on *Halimeda*. III. A daily cycle of chloroplast migration within segments. *Bot. Marina* 33:31–45.

Dreyer, E. M., and Weisenseel, M. H. 1979. Phytochrome-mediated uptake of calcium in *Mougeotia* cells. *Planta* 146:31–39.

Dring, M. J. 1988. Photocontrol of development in algae. *Ann. Rev. Plant Physiol. Mol. Bio.* 39:157–174.

Ekelund, N. G. A., Sundqvist, C., Quail, P., and Vierstra, R. D. 1985. Chromophore rotation in 124-kdalton *Avena sativa* phytochrome as measured by light-induced changes in linear dichroism. *Photochem. Photobiol.* 41:221–223.

Etzold, H. 1965. Der Polarotropismus und Phototropismus der Chloronemen von *Dryopteris filix mas* L. Schott. *Planta* 64:254–280.

Franke, W., VanDerWoude, W., and Morré, D. 1972. Tubular and filamentous structures in pollen tubes: Possible involvement as guide elements in protoplasmic streaming and vectorial migration of secretory vesicles. *Planta* 105:317–341.

Furuya, M. 1987. *Phytochrome and Photoregulation in Plants*. Academic Press, Tokyo.

Furuya, M. 1989. Molecular properties and biogenesis of phytochrome I and II. *Adv. Biophys.* 25:133–167.

Gabryś, H., Walczak, T., and Haupt, W. 1984. Blue-light-induced chloroplast orientation in *Mougeotia*. Evidence for a separate sensor pigment besides phytochrome. *Planta* 160:21–24.

Galway, M. E., and Hardham, A. R. 1986. Microtubule reorganization, cell wall synthesis and establishment of the axis of elongation in regenerating protoplasts of the alga *Mougeotia*. *Protoplasma* 135:130–143.

Grolig, F. 1986. Calcium-Vesikel und lichtabhängige Chloroplastenreorientierung bei *Mougeotia* spec. Ph.D. Thesis, Universität Giessen, Federal Republic of Germany.

Grolig, F., and Wagner, G. 1987. Vital staining permits isolation of calcium vesicles from the green alga *Mougeotia*. *Planta* 171:433–437.

Grolig, F., and Wagner, G. 1988. Light-dependent chloroplast reorientation in *Mougeotia* and *Mesotaenium*: Biased by pigment-regulated plasmalemma anchorage sites to actin filaments? *Bot. Acta* 101:2–6.

Grolig, F., and Wagner, G. 1989. Characterization of the isolated calcium-binding vesicles from the green alga *Mougeotia scalaris*, and their relevance to chloroplast movement. *Planta* 177:169–177.

Grolig, F., Weigang-Köhler, K., and Wagner, G. 1990. Different extent of F-actin bundling in walled cells and in protoplasts of *Mougeotia scalaris*. *Protoplasma*, 157: 225–230.

Hartmann, K. M. 1983. Action spectroscopy. In W. Hoppe, W. Lohmann, H. Markl, and H. Ziegler (eds.), *Biophysics*, Springer, Berlin, pp. 115–144.

Hartwig, R., Schweiger, M., Schweiger, R., and Schweiger, H. G. 1985. Identification of a high molecular weight polypeptide that may be part of the circadian clockwork in *Acetabularia*. *Proc. Natl. Acad. Sci. U.S.A.* 82:6899–6902.

Hartwig, R., Schweiger, R., and Schweiger, H.-G. 1986. Circadian rhythm of the synthesis of a high molecular weight protein in anucleate cells of the green alga *Acetabularia*. *Eur. J. Cell Biol.* 41:139–141.

Haupt, W. 1982. Light-mediated movements of chloroplasts. *Ann. Rev. Plant Physiol.* 33:205–233.

Haupt, W. 1983a. Movement of chloroplasts under the control of light. In F. Round and G. Chapman (eds.), *Progress in Phycological Research*, Vol. 2. Elsevier, Amsterdam, pp. 227–281.

Haupt, W. 1983b. The perception of light direction and orientation responses in chloroplasts. In D. J. Cosens and D. Vince-Price (eds.), *The Biology of Photoperception*. Society for Experimental Biology, Great Britain, pp. 423–442.

Haupt, W. 1987. Phytochrome control of intracellular movement. In M. Furuya (ed.) *Phytochrome and Photoregulation in Plants*. Acadmic Press, Tokyo, pp. 225–237.

Jacobshagen, S., Altmüller, D., Grolig, F., and Wagner, G. 1986. Calcium pools, calmodulin and light-regulated chloroplast movements in *Mougeotia* and *Mesotaenium*. In A. J. Trewavas (ed.), *Molecular and Cellular Aspects of Calcium in Plant Development*. Plenum Press, New York, pp. 201–217.

Kadota, A., Inoue, Y., and Furuya, M. 1986. Dichroic orientation of phytochrome intermediates in the pathway from Pr to Pfr as analyzed by double flash irradiations in polarotropism of *Adiantum* protonemata. *Plant Cell Physiol.* 27:867–873.

Kadota, A., Koyama, M., Wada, M., and Furuya, M. 1984. Action spectra for polarotropism and phototropism in protonemata of the fern *Adiantum capillusveneris*. *Physiol. Plant.* 61:327–330.

Kamiya, N. 1986. Cytoplasmic streaming in giant algal cells: A historical survey of experimental approaches. *Bot. Mag. Tokyo* 99:441–467.

Kataoka, H. 1975a. Phototropism in *Vaucheria geminata*. I. The action spectrum. *Plant Cell Physiol.* 16:427–437.

Kataoka, H. 1975b. Phototropism in *Vaucheria geminata*. II. The mechanism of bending and branching. *Plant Cell Physiol.* 16:439–448.

Kataoka, H. 1981. Expansion of *Vaucheria* cell apex caused by blue or red light. *Plant Cell Physiol.* 22:583–595.

Kataoka, H., and Weisenseel, M.H. 1988. Blue light promotes ionic current influx at the growing apex of *Vaucheria terrestris*. *Planta* 173:490–499.

Kicherer, R. M. 1985. Endogene and Blaulicht-induzierte Ionenströme bei der Alge *Vaucheria sessilis*. Ph.D. Thesis, Univeristät Erlangen-Nürnberg, Federal Republic of Germany.

Kidd, D. G., and Lagarias, J. C. 1990. Phytochrome from the green alga *Mesotaenium caldariorum*. Purification and preliminary characterization. *J. Biol. Chem.* 265:7029–7035.

Klein, K., Wagner, G., and Blatt, M. R. 1980. Heavy-meromyosin-decoration of microfilaments from *Mougeotia* protoplasts. *Planta* 150:354–356.

Koop, H.-U., and Kiermayer, O. 1980a. Protoplasmic streaming in the giant unicellular green alga *Acetabularia mediterranea* I. Formation of intracellular transport systems in the course of cell differentiation. *Protoplasma* 102:147–166.

Koop, H.-U., and Kiermayer, O. 1980b. Protoplasmic streaming in the giant unicellular alga *Acetabularia mediterranea*. II. Differential sensitivity of movement systems to substances acting on microfilaments and microtubules. *Protoplasma* 102:295–306.

Koop, H.-U., Schmid, R., Heunert, H.-H., and Milthaler, B. 1978. Chloroplast migration: A new circadian rhythm in *Acetabularia*. *Protoplasma* 97:301–310.

Kraml, M., Büttner, G., Haupt, W., and Herrmann, H. 1988. Chloroplast orientation in *Mesotaenium*: The phytochrome effects strongly potentiated by interaction with blue light. *Protoplasma (Suppl. 1)*:172–179.

Lechowski, Z., and Bialczyk, J. 1988. Action spectrum for interaction between visible and far-red light on face chloroplast orientation in *Mougeotia*. *Plant Physiol.* 88:189–193.

Lew, R. R., Serlin, B. S., Schauf, C.L., and Stockton, M. E. 1990a. Red light regulates calcium-activated potassium channels in *Mougeotia* plasma membrane. *Plant Physiol.* 92:822–830.

Lew, R. R., Serlin, B. S., Schauf, C. L., and Stockton, M. E. 1990b. Calcium activation of *Mougeotia* potassium channels. *Plant Physiol.* 92:831–836.

Lopez-Figueroa, F., and Niell, F. X. 1989. A possible control by a phytochrome-like photoreceptor of chlorophyll synthesis in the green alga *Ulva rigida*. *Photochem. Photobiol.* 50:263–266.

Lopez-Figueroa, F., Lindemann, P., Braslavsky, S. E., Schaffner, K., Schneider-Poetsch, H. A. W., and Rüdiger, W. 1989. Detection of a phytochrome-like protein in macroalgae. *Bot. Acta* 102:178–180.

Maekawa, T., and Nagai, R. 1988. Reorganization of microtubule bundles in *Dichotomosiphon*: Its implications in the light-induced translocation of cytoplasm. *Protoplasma* (Suppl. 1):162–171.

Maekawa, T., Tsutsui, I., and Nagai, R. 1986. Light-regulated translocation of cytplasm in green alga *Dichotomosiphon*. *Plant Cell Physiol.* 27:837–851.

Marchant, H. J. 1976. Actin in the green algae *Coleochaete* and *Mougeotia*. *Planta* 131:119–120.

Menzel, D. 1986. Visualization of cytoskeletal changes through the life cycle in *Acetabularia*. *Protoplasma* 134:30–42.

Menzel, D. 1987. The cytoskeleton of the giant coenocytic green alga *Caulerpa* visualized by immunocytochemistry. *Protoplasma* 139:71–76.

Menzel, D. 1988. How do giant plant cells cope with injury?—The wound response in siphonous green algae. *Protoplasma* 144:73–91.

Menzel, D., and Elsner-Menzel, C. 1989a. Co-localization of particle transport with microtubules in cytoplasmic exudates of the siphonous green alga *Bryopsis*. *Bot. Acta* 102:241–248.

Menzel, D., and Elsner-Menzel, C. 1989b. Actin-based chloroplast rearrangements in the cortex of the giant coenocytic green alga *Caulerpa*. *Protoplasma* 150:1–8.

Menzel, D., and Elsner-Menzel, C. 1989c. Maintenance and dynamic changes of cytoplasmic organization controlled by cytoskeletal assemblies in *Acetabularia* (Chlorophyceae). In Coleman, A. W., Goff, L. J. and Stein-Taylor, J. R. (eds.) *Algae as Experimental Systems*. Alan R. Liss, New York, pp. 71–91.

Menzel, D., and Schliwa, M. 1986a. Motility in the siphonous green alga *Bryopsis*. I. Spatial organization of the cytoskeleton and organelle movements. *Eur. J. Cell Biol.* 40:275–285.

Menzel, D., and Schliwa, M. 1986b. Motility in the siphonous green alga *Bryopsis*. II. Chloroplast movement requires organized arrays of both microtubules and actin filaments. *Eur. J. Cell Biol.* 40:286–295.

Mohr, H., and Schopfer, P. 1978. Lehrbuch der Pflanzenphysiologie. Springer-Verlag, Berlin.

Nagai, R., and Fukui, S. 1981. Differential treatment of *Acetabularia* with cytochalasin B and N-ethylmaleimide with special reference to their effects on cytoplasmic streaming. *Protoplasma* 109:79–89.

Neuscheler-Wirth, H. 1970. Photomorphogenese und Phototropismus bei *Mougeotia*. Z. *Pflanzenphysiol.* 63:238–260.

Nultsch, W., Pfau, J., and Ruffer, U. 1981. Do correlations exist between chromatophore arrangement and photosynthetic activity in seaweeds? *Mar. Biol.* 62:111–117.

Quail, P. H. 1984. Phytochrome: A regulatory photoreceptor that controls the expression of its own gene. *Trends Biochem. Sci.* 9:450–453.

Roberts, D. M. 1989. Detection of a calcium-activated protein kinase in *Mougeotia* by using synthetic peptide substrates. *Physiol. Plant.* 91:1613–1619.

Roux, S. J. 1984. Ca^{2+} and phytochrome action in plants. *Bioscience* 35:25–29.

Rüdiger, W. 1980. Phytochrome, a light receptor of plant photomorphogenesis. In J. D. Dunitz et al. (eds.), *Structure and Bonding*, Vol. 40. Springer, Berlin, pp. 101–140.

Russ, U., Grolig, F., and Wagner, G. 1991. Changes of cytoplasmic free Ca^{2+} in the green alga *Mougeotia scalaris* as monitored with indo-1, and their effect on the velocity of chloroplast movements. *Planta*, 184:105–112.

Russ, U., Grolig, F., and Wagner, G. 1988. Differentially adsorbed vital dyes inhibit chloroplast movement in *Mougeotia scalaris*. *Protoplasma* (*Supp. 1*): 180–184.

Schönbohm, E. 1973. Die lichtinduzierte Verankerung der Plastiden im cytoplasmatischen Wandbelag: Eine phytochromgesteuerte Kurzzeitreaktion. *Ber. Deutsch. Bot. Ges.* 86:423–430.

Schönbohm, E. 1980. Phytochrome and non-phytochrome dependent blue light effects in intracellular movements in fresh-water algae. In H. Senger (ed.), *The Blue Light Syndrome*, Springer, Berlin, pp. 69–95.

Schönbohm, E., and Hellwig, H. 1979. Zum Photorezeptor-Problem der Schwachlichtbewegung des *Mougeotia* Chloroplasten im Blau, bzw. Hellrot bei niederen Temperaturen. *Ber. Deutsch. Bot. Ges.* 92:749–762.

Schönbohm, E., Meyer-Wegener, J., and Schönbohm, E. 1990. No evidence for Ca^{2+} influx as an essential link in the signal transduction chains of either light-oriented chloroplast movements or Pfr-mediated chloroplast anchorage in *Mougeotia J. Photochem. Photobiol.* 5:331–341.

Schönbohm, E., and Schönbohm, E. 1984. Biophenole: Steuernde Faktoren bei der lichtorientierten Chloroplastenbewegung? *Biochem. Physiol. Pflanz.* 179:498–505.

Senn, G. 1908. Die Gestalt- und Lageveränderung der Pflanzen-Chromatophoren. Engelmann, Leipzig.

Serlin, B. S., and Ferrell, S. 1989. The involvement of microtubules in chloroplast rotation in the alga *Mougeotia*. *Plant Sci.* 60:1–8.

Serlin, B. S., and Roux, S. J. 1984. Modulation of chloroplast movement in the green alga *Mougeotia* by the Ca^{2+}-ionophore, A 23187, and by calmodulin antagonists. *Proc. Natl. Acad. Sci. U.S.A.* 81:6368–6372.

Sharrock, R. A., and Quail, P. H. 1989. Novel phytochrome sequences in *Arabidopsis thaliana*: Structure, evolution, and differential expression of a plant regulatory photoreceptor family. *Genes Dev.* 3:1745–1757.

Sugimoto, T., Ito, E., and Suzuki, H. 1987. Interpretation of the "dichroic orientation" of phytochrome. *Photochem. Photobiol.* 46:517–523.

Sundqvist, C., and Björn, L. O. 1983a. Light-induced linear dichroism in photoreversibly photochromic sensor pigments. II. Chromophore rotation in immobilized phytochrome. *Photochem. Photobiol.* 37:69–75.

Sundqvist, C., and Björn, L. O. 1983b. Light-induced linear dichroism in photoreversibly photochromic sensor pigments. III. Chromophore rotation estimated by polarized light reversal of dichroism. *Physiol. Plant.* 59:263–269.

Thimann, K., and Curry, G. 1961. Phototropism. In W. McElroy, and B. Glass (eds.), *Light and life*. John Hopkins Press, Baltimore, pp. 646–669.

Wada, M., and Kadota, A. 1989. Photomorphogenesis in lower green plants. *Ann. Rev. Plant Physiol. Mol. Bio.* 40:169–191.

Wagner, G., and Bellini, E. 1976. Light-dependent fluxes and compartmentation of calcium in the green alga *Mougeotia*. *Z. Pflanzenphysiol.* 79:283–291.

Wagner, G., and Grolig, F. 1985. Molecular mechanisms of photoinduced chloroplast movements. In G. Colombetti, F. Lenci, and P.-S. Song (eds.), *Sensory Perception and Transduction in Aneural Organisms*, Plenum Press, New York, pp. 281–298.

Wagner, G., Grolig, F., and Altmüller, D. 1987. Transduction chain of low irradiance response of chloroplast reorientation in *Mougeotia* in blue or red light. *Photobiochem. Photobiophys.* (Suppl.) 183–189.

Wagner, G., and Klein, K. 1978. Differential effect of calcium on chloroplast movement in *Mougeotia*. *Photochem. Photobiol.* 27:137–140.

Wagner, G., and Klein, K. 1981. Mechanism of chloroplast movement in *Mougeotia*. *Protoplasma* 109:169–185.

Wagner, G., and Rossbacher, R. 1980. X-ray microanalysis and chlorotetracycline staining of calcium vesicles in the green alga *Mougeotia*. *Planta* 149:298–305.

Wagner, G., Valentin, P., Dieter, P., and Marmé, D. 1984. Identification of calmodulin in the green alga *Mougeotia* and its possible function in chloroplast reorientational movement. *Planta* 162:62–67.

Weidinger, M., and Ruppel, H. G. 1985. Ca^{2+}-requirement for a blue-light-induced chloroplast translocation in *Eremosphaera viridis*. *Protoplasma* 124:184–187.

Weisenseel, M. 1968. Vergleichende Untersuchungen zum Einfluß der Temperatur auf die lichtinduzierte Chloroplastenverlagerung. *Z. Pflanzenphysiol.* 59:56–69.

Williamson, R. E. 1975. Cytoplasmic streaming in *Chara*: A cell model activated by ATP and inhibited by cytochalasin B. *J. Cell Sci.* 17:655–668.

Williamson, R. E. 1981. Free Ca^{2+} concentration in the cytoplasm: A regulator of plant cell function. *What's New in Plant Physiology* 12:45–48.

Yatsuhashi, H., Wada, M., and Hashimoto, T. 1987. Dichroic orientation of phyto-
    chrome and blue-light photoreceptor in *Adiantum* protonemata as determined by
    chloroplast movement. *Acta Physiol. Plant* 9:163–173.
Zurzycki, J. 1955. Chloroplasts arrangement as a factor in photosynthesis. *Acta Soc.
    Bot. Pol.* 24:27–63.

Chapter 3

Cytoplasmic Streaming in Characean Algae: Mechanism, Regulation by Ca^{2+}, and Organization

Richard E. Williamson

Introduction

The endoplasm of the internodal cells of the characean algae streams at more than 50 μm s^{-1}. The giant size and simple organization of these cells have for over 200 years made them favorite objects for studying motility by microscopy and ingenious experimentation (reviewed by Kamiya 1959, 1960, 1962, 1981). Studies over the last 25 years have identified the structures causing streaming and some of the proteins that they contain and have begun to elucidate their function, organization, and regulation.

General Aspects of Cellular Organization

The basically simple organization of characean internodal cells contributes importantly to their value for experimentation. Cells grow as right cylinders with lengths of several centimeters and diameters of 1 mm or more in the large-celled, experimentally favored species such as *Chara corallina* and *Nitella flexilis*. The internodal cells are separated by clusters of much smaller nodal cells, the two cell types being the products of alternate cells from divisions occurring behind the apical cell.

The organization of the cytoplasm is shown diagrammatically in Figure

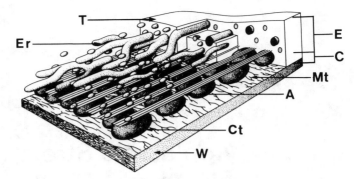

Figure 3.1. Diagramatic cutaway view of a small part of the wall and peripheral cytoplasm of a characean internodal cell showing the spatial relationships between some of the major cell components. The cortical cytoplasm (C) containing the microtubules (Mt), chloroplasts (Ct), and actin bundles (A) is stationary whereas the endoplasm (E), containing endoplasmic reticulum (Er) and many small spherical to ovoid organelles, streams parallel to the actin bundles at 50–100 μm s^{-1}. T = tonoplast; W = wall.

3.1. The large central vacuole occupying some 95% of the cell volume is surrounded by a cylinder of cytoplasm, typically 5–10 μm thick but capable of varying markedly. In living cells, the layer of stationary cortical cytoplasm is readily distinguished from the layer of streaming endoplasm that lies within it. The endoplasm is thus bounded at its inner surface by the tonoplast, and the cortical cytoplasm is bounded at its outer surface by the plasma membrane. The boundary between stationary cortex and endoplasm lies approximately at the vacuole-facing surface of the chloroplasts which are arranged in helically wound files in the cortex. The cortex also contains representatives of most other categories of cellular organelles with the exception of nuclei and small vacuoles. The endoplasm has few chloroplasts but many nuclei and other organelles (Pickett-Heaps, 1975). A sector occupying half the cell's circumference is underlain with endoplasm streaming in one direction and is separated by narrow zones lacking chloroplasts (the neutral line or indifferent zone) from cytoplasm streaming in the opposite direction and covering the other half of the cell's circumference. The helical chloroplast files run parallel to the direction of streaming.

Foundations for Recent Studies of Streaming

Many studies made between about 1955 and 1975 underpin recent research and are described in more detail by Kamiya (1986). I will mention just some of the more important. Kamiya and Kuroda (1956) deduced that the motive force for streaming is generated at the interface between cortex and endoplasm by measuring the velocity profile across cell fragments lacking a vacuole. The magnitude of this motive force was estimated by determining the force required to stop streaming using centrifugal (Kamiya and Kuroda, 1958) and hydrodynamic (Tazawa, 1968) forces. Hayashi (1964) and others showed that locally damaging the cortex impairs streaming of the endoplasm beneath it, and Kamitsubo (1966) demonstrated fibers whose disappearance and regrowth correlate with the ability of a region of cortical cytoplasm to support endoplasmic streaming. During the recovery phase, the long, straight fibers support the movements of organelles in their vicinity, whereas organelles a few micrometers distant remain immobile. Nagai and Rebhun (1966) and Pickett-Heaps (1967) showed that these fibers are bundles of 6-nm filaments. When dislodged from their anchorage to the cortex, the fibers show independent movement (Kamitsubo, 1972a), thus providing the likely origin for at least some of the moving fibers seen in fragments of cytoplasm mechanically expelled from cut cells (Jarosch, 1956).

The current era of streaming research can most conveniently be dated from two findings in the mid-1970s: the demonstration that characean cells contain (as long suspected) filaments of actin (Palevitz et al., 1974; Williamson, 1974), the major protein of muscle thin filaments, and the demonstration that cytoplasmic streaming can be supported by ATP after the tonoplast has been removed or permeabilized (Williamson, 1975; Tazawa et al., 1976).

Mechanism of Streaming

Localization and Organization of Cytoskeletal Proteins

Actin. Actin, a filamentous polymer of a 42,000-M_r subunit, is ubiquitous in eukaryotic cells but is perhaps most familiar as the major protein of muscle thin filaments. It binds to myosin, the major protein of muscle thick filaments, activating its ATPase probably by stimulating product release (Pollard and Cooper, 1986). Actin has been localized in characean algae by its reaction with muscle myosin fragments (Palevitz and Hepler, 1975) and with fluorochrome-derivatized antibodies (Fig. 3.2a; Williamson and Toh, 1979; Owaribe et al., 1979), phallotoxins, myosin fragments (Nothnagel et al., 1981), and the tropomyosin–troponin complex of skeletal muscle (Shimmen and Yano, 1986). All reactions point to actin being the major component of the 6-nm filament bundles adjacent to the vacuole-facing surface of the cortical chloroplasts. Whether filamentous actin exists elsewhere in the cell is not adequately addressed by these studies, which used perfused cells containing limited quantities of endoplasm and possibly partially extracted cortical cytoplasm. Grolig et al. (1988) found no other locations of actin when carrying out immunofluorescence on "whole" cells that contain extensive endoplasm (in practice, some endoplasmic components will inevitably be

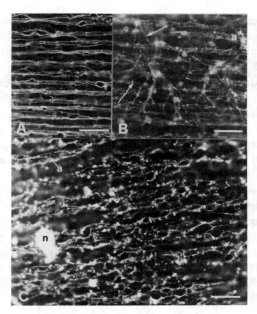

Figure 3.2. Immunofluorescent localization of actin (A) and myosin (B, C) in *Chara* internodal cells. (A) Actin is confined to the filament bundles. (B and C) Myosin occurs in the filament bundles (a), on beaded strands (arrows) that are probably endoplasmic reticulum cisternae, on numerous small organelles and on nuclei (n). Bars = 20μm. Reproduced from Grolig *et al.*, 1988.

lost during chemical fixation and processing of these fragile cells), and Williamson et al. (1987) argued from immunoblots and selective extraction that the cortex contains little actin besides that in the known filament bundles. The actin of internodal cells is resolved as a single isoform on two-dimensional gels (Williamson et al., 1987).

The actin filament bundles are closely associated with and may be physically linked by cross-bridges (McLean and Juniper, 1988) to the outer envelope membrane of the chloroplasts (Nagai and Rebhun, 1966; Pickett-Heaps, 1967; Palevitz and Hepler, 1975). The chloroplasts, however, can be exhaustively extracted with detergent without releasing the actin bundles (Williamson, 1985). Branches of the actin bundles that pass between the chloroplasts and run for some distance beneath the plasma membrane may provide anchorage that survives chloroplast extraction (Williamson et al., 1986). The bundled actin filaments show hexagonal packing (Kamiya and Nagai, 1982) without any indications of discontinuities along the length of the bundles.

All the filaments in a single bundle have the same polarity, which is revealed visually when they react with fragments of muscle myosin (Palevitz et al., 1974). The myosin fragments form structures resembling arrowheads, and the arrowheads point in the direction opposite to that in which the endoplasm around the bundle is streaming (Kersey et al., 1976). A myosin reacting with these actin cables can be predicted from the geometry of a muscle sarcomere (Huxley, 1963) to move in the same direction as the endoplasm streams. This polarity is accompanied by different assembly characteristics at the two ends of each filament. The barbed or plus end extends much more rapidly by monomer addition than does the pointed or minus end of the filament, but monomer addition can also be regulated by proteins binding specifically to the plus or the minus end (Pollard and Cooper, 1986).

Myosin. Much less is known of myosin than of actin because it is difficult to recognize electron-microscopically, and only antibodies are available to localize it. Because sequence conservation of myosins is much less than that in different actins, few antibodies to animal myosins cross-react widely even within the animal kingdom. Kato and Tonomura (1977) partially purified small quantities of a *Nitella* myosin whose ATPase is activated by muscle actin and that forms small aggregates visible in the electron microscope. Grolig et al. (1988) showed that a monoclonal antibody to a mouse, nonmuscle myosin reacts on immunoblots with a polypeptide migrating (like that of Kato and Tonomura's myosin) close to the heavy chain of rabbit skeletal muscle myosin (about 200,000 M_r). In addition, however, the same antibody recognizes a polypeptide of 110,000 M_r that does not seem to be a proteolysis product of the larger polypeptide. The reaction with the smaller polypeptide raises (but does not settle) the interesting possibility that *Chara* has two forms of myosin. These may equate with the myosin I and myosin II categories. Myosin II's are the large myosins familiar from muscle and many nonmuscle cells that have two enzymically active heads and an elongated tail that can aggregate them into bipolar filaments. Myosin I's are the much smaller molecules (110,000–140,000 M_r) known in mammals, *Dictyostelium* and *Acanthamoeba* that have a single head and, lacking the long tail of myosin I's, cannot aggregate (reviewed in Korn and Hammer, 1988; Titus et al., 1989). An affinity for lipids is thought to bind them to organelle and cell membranes (Adams and Pollard, 1989). Sequence homologies between myosin I's and the head region of myosin II's could allow antibodies to recognize both categories.

Immunofluorescence using the antimyosin monoclonal (Grolig et al., 1988) showed that the proteins it recognizes are distributed continuously along the bundles of actin filaments and on the surface of the motile organelles of the endoplasm (see Fig. 3.2b). They are absent from the chloroplast envelope surface and from the surface of other nonmotile organelles in the cortex. Grolig et al. also described beaded endoplasmic strands that react with the antibody and ramify through the endoplasm. These seem likely to be endoplasmic reticulum, which has been shown in other cells to undergo a similar beading response with hypotonic solutions (Dabora and Sheetz, 1988) and is known to be abundant in characean endoplasm (Bradley, 1973; Kachar and Reese, 1988). Electron microscopy had previously generated two speculations regarding the location of myosin in the endoplasm: Nagai and Hayama (1979) proposed that myosin occurred on membranous protuberances from an unidentified type of endoplasmic organelle, whereas Williamson (1979) found large, detergent-resistant filaments that were thought to be associated with the endoplasmic reticulum. Neither type of structure has been further characterized in subsequent studies by, for example, immunoelectron microscopy.

Microtubules. The second major cytoskeletal system of plant cells consists of microtubules, polymers of α- and β-tubulin. The electron microscopy that identified bundles of filaments at the cortex–endoplasm boundary was particularly influential because it also showed that microtubules were confined to the region between the chloroplasts and the plasma membrane from where they could not contribute force to move the endoplasm on the vacuolar side of the chloroplasts (Nagai and Rebhun, 1966; Pickett-Heaps, 1967). This neat separation of microtubules and actin has recently been undermined with the demonstration by Wasteneys (1988) that immunofluorescence detects microtubules in the endoplasm in addition to the familiar cortical array seen by electron microscopy. The position and orientation of some of the microtubules suggest that they are associated with the bundles of actin filaments. Others deeper in the endoplasm are not aligned parallel to the streaming with current preparative methods but may be associated with the numerous nuclei traveling in the endoplasm.

Experimental Studies of the Mechanism of Streaming

Whole Cells. The use of inhibitors on whole cells was influential in formulating the view that actin filaments and not microtubules are involved in streaming, since cytochalasin B, which affects actin filaments (MacLean-Fletcher and Pollard, 1980; Pollard and Mooseker, 1981), inhibits streaming (Wessels et al., 1971; Williamson, 1972; Bradley, 1973), but colchicine (Bradley, 1973; see also Pickett-Heaps, 1967) and other microtubule-depolymerizing agents (Wasteneys and Williamson, unpublished) do not. No inhibitors specific for myosin are available, but a number of studies have employed N-ethylmaleimide, a sulfhydryl reagent that covalently binds to and inhibits enzymes such as myosins that have essential sulfhydryl groups. Chen and Kamiya (1975) extended earlier studies by displacing characean endoplasm to one end of the cell with centrifugation (actin remains in situ) and applying the inhibitor to either the endoplasm-rich or endoplasm-poor end of the cell. The endo-plasm could then be displaced by a second centrifugation to bring it into contact with the actin at the other end of the cell that had received a different treatment. They showed with such experiments that N-ethylmaleimide acts on the endoplasm, not on the cortex, which is consistent with an endoplasmic myosin. The experiments in which motility supported by exogenous ATP is inhibited by N-ethylmaleimide and restored with muscle myosin fragments (Kuroda and Kamiya, 1975; discussed later) increase confidence that the inhibitor is not simply slowing metabolism. Cytochalasin B shows the reverse specificity in such centrifugation, localized-inhibitor experiments, inhibiting the actin-containing cortical cytoplasm but not the endoplasm (Nagai and Kamiya, 1977). The ready displacement of endoplasm in cytochalasin-inhib-ited cells (Kuroda and Kamiya, 1981) is consistent with cytochalasin inhib-iting motive force generation rather than increasing endoplasmic viscosity. Kamitsubo et al. (1989) noted that the endoplasm continues to stream parallel to the helically wound actin bundles rather than being displaced longitudinally even by moderately severe centrifugation (900 × g). They argue that this provides physical evidence for connections between the actin bundles and the endoplasm.

Membrane-Permeabilized Cells and Cytoplasmic Fragments. Work on the mechanism and regulation of streaming has been greatly facilitated by tech-niques to remove or permeabilize either the tonoplast or plasma membrane so that organelle movements can be sustained with exogenous ATP and/or macromolecules such as antibodies introduced. Intracellular perfusion was introduced by Kamiya and Kuroda in 1955 (see Kamiya and Tazawa, 1966; Kamiya, 1986) and was simplified by Tazawa (1964). It involves reducing the cell's turgor pressure and carefully cutting off the ends of the cell, so that solution can flow through the interior under a small gradient of hydrostatic pressure. Using gentle perfusion with solutions broadly resembling the natu-ral vacuolar sap, perfusion was used to study vacuolar ion transport, to measure the motive force for streaming, and so on. Its value was extended by removing the tonoplast either by sweeping it away using a rapid flow of solution (Fig. 3.3a; Williamson, 1975) or by lowering the normally millimo-lar vacuolar Ca^{2+} concentration by chelation (Tazawa et al., 1976). In both cases, judicious control of solution composition (reviewed later under "Reg-ulation at the Action Potential") allows the remaining cytoplasm to show ATP-dependent streaming at rates equaling those seen in vivo. The particular beauty of the characean cell is the robustness of the cortical cytoplasm, including the actin cables, which remains in situ, so facilitating solution changes involving both macromolecules and small solutes.

First removing ATP and then restoring it by perfusion significantly clarifies the interaction between endoplasmic organelles and the actin bundles (Wil-

80   Williamson

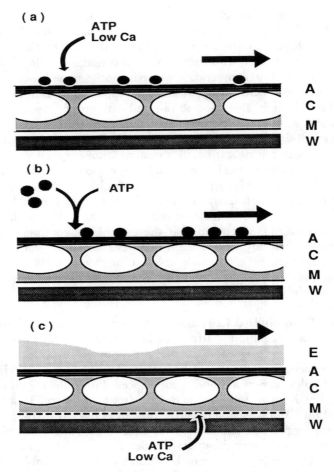

Figure 3.3. Reactivation and reconstitution of motility using characean cells. (a). Rapid perfusion removes the tonoplast and many cytoplasmic components; motility is moderately but irreversibly Ca^{2+}-sensitive. (b) Motility reconstituted by the addition of μm-sized beads coated with exogenous myosin is Ca^{2+}-sensitive only if the added myosin is Ca^{2+}-sensitive. (c) Plasma membrane permeabilization extracts few macromolecules and motility is highly and reversibly sensitive to external Ca^{2+}. E = endoplasm, A = actin, C = chloroplasts, M = plasma membrane, W = wall.

liamson, 1975). In the absence of ATP, organelles are firmly bound to the actin cables in a manner consistent with a rigor complex formed between myosin-coated organelles and actin filaments. When ATP is restored, organelles move along the actin cables at rates equaling those seen in vivo and following all deviations in the path of the actin filaments. The cables themselves do not move and in particular show no wave propagation. Organelles readily detach from the cables in the presence of ATP and then show only Brownian-type motion until recontacting the actin. Larger fragments of cytoplasm can move as a unit, however, including organelles that are unlikely to be in direct contact with the actin cables. Presumably, viscous coupling of some sort is involved. Many of these observations were confirmed by Kachar (1985), who expelled cytoplasm from cut cells and observed movements along disordered actin cables on a slide–coverslip preparation. In these preparations, Kachar showed by video analysis that the velocity distribution of the moving organelles is biphasic. Small spherical organelles showed slower movements (11 μm s^{-1}) that accounted for 99 percent of the total movements, whereas the rare movements at in vivo speeds involved "faint tubular structures" that could change shape during movement.

Other studies of motility have involved demembranating the cytoplasmic fragments that can be squeezed from cut cells or allowed to flow out while a negative pressure is applied to the cell. Such fragments are bounded by a surface membrane that is derived from the tonoplast (Sakano and Tazawa, 1986). Kuroda and Kamiya (1975) permeabilized such fragments by Ca^{2+} chelation and mechanical puncture and used exogenous ATP to sustain the rotation of chloroplasts, a phenomenon that probably results from chloroplasts displaced from the cortical cytoplasm remaining associated with their unipolar actin cables. Force generation by the actin cables will cause movement of the chloroplast once it is not restrained by its normal anchorage system in the cell. Higashi-Fujime (1980) showed that chains of chloroplasts linked by actin cables exhibit ATP-dependent movements in vitro and that freed bundles of actin filaments would show translational movements or rotate as loops. Higashi-Fujime speculated that the myosin presumed to be required for these movements may have been immobilized on the glass slide, a proposal that gains plausibility from the recent success in causing purified actin to move over myosin adsorbed to the glass slide (Kron and Spudich, 1986). The rotating rings are presumably the in vitro counterpart of some of the rings and polygons seen by Jarosch (1956) and others subsequently.

Reconstituting Motility. The components required for motility in the membrane-permeabilized preparations that have been described can be analyzed by recombining cytoplasmic fractions or pure proteins to reconstitute motility. Shimmen and Tazawa (1982) reintroduced endoplasm expelled from *Chara* into an endoplasm-free cell of *Nitella* in which any residual myosin had been inhibited with N-ethylmaleimide (see Fig. 3.3b). The organelles were prepared for transfer by pelleting a low-speed supernatant at 14,000 g for 5 to 10 min and resuspending the endoplasmic "granules" in a medium compatible with reactivating streaming in perfused cells. The organelles moved with a polarity specified by the actin cables on which they were placed and at velocities of up to 50 percent of that of the in vivo streaming.

The organelles clearly retained the myosin required for movement, but the tightness of its binding to the organelles has not been fully explored by this method. Subsequently, Kohno and Shimmen (1988) transplanted motile organelles from pollen tubes onto the actin cables of characean internodal cells and showed that they move almost as rapidly as do characean organelles and much more rapidly than in situ organelles in pollen tubes. The authors suggest that this may reflect the larger size of the characean actin bundles or a reduced resistance in the reconstituted system compared to the pollen tube. Organelles from *Acanthamoeba* translocate at <0.1 μm s^{-1} along characean actin cables (Adams and Pollard, 1986) and the movement is inhibited by antibodies to myosin I but not by antibodies to myosin II. This suggests that some myosin I is organelle-associated in *Acanthamoeba* and shows that it can transduce chemical energy into mechanical energy.

Kamiya and Kuroda (1975) pioneered the use of purified proteins to reconstitute motility by showing that subfragments of muscle myosin restore chloroplast rotation (albeit very slowly) in demembranated cytoplasmic fragments. N-ethylmaleimide was used to inhibit rotation, a process that, from this inhibition-rescue experiment, can be deduced to depend on myosin. Kuroda (1983) similarly had some limited success in using myosin subfragments to restore movement to similarly inactivated characean endoplasm in cut-open cells. This approach was, however, given a major impetus by the development of much more convenient techniques with which to apply exogenous myosin to longitudinally opened cells (Sheetz and Spudich, 1983) or to perfused cells (Shimmen and Yano, 1984). Myosin II (Sheetz and Spudich 1983, Shimmen and Yano 1984) or myosin I (Albanesi et al., 1985; Mooseker and Coleman, 1989) is applied covalently coupled or physically adsorbed to small, organelle-sized beads. These move along the actin cables at velocities ranging from <0.01 μm s^{-1} (brush border myosin I) up to about 6 μm s^{-1} (rabbit skeletal muscle myosin II; reviewed by Shimmen, 1988). Such experiments will be discussed further in the context of the regulation of characean actin and myosin by Ca^{2+}. In terms of the mechanism of streaming, however, they show that organelle-associated myosins could indeed be expected to generate movement but that, with the myosins so far tried, the rates of bead movement are well below those shown by characean or even transplanted pollen tube organelles. The rate of movement of beads coated with *Acanthamoeba* myosin I (Albanesi et al., 1985) is also less than the velocities shown by *Acanthamoeba* organelles undergoing myosin I-dependent movement (Adams and Pollard, 1986).

Mechanism of Cytoplasmic Streaming

A substantial body of evidence is consistent with streaming resulting from an endoplasmic myosin that can associate with endoplasmic organelles and move along stationary trackways of unipolar actin filaments. Although simple in outline, many uncertainties remain. These include the number of myosins and their aggregation state; the nature of myosin's association with endoplasmic organelles; the molecular or organizational properties of plant myosins that allow plant organelles to travel so much faster than *Acanthamoeba* organelles and than beads coated with animal myosins; and whether individual organelles reacting with actin contribute significantly to moving the bulk of the endoplasm, parts of which can be several micrometers from the actin.

Of these questions, only the latter has been the subject of substantial analysis. Nothnagel and Webb (1982) concluded that neither individual myosin molecules nor myosin-coated organelles traveling along the actin could exert enough viscous pull to move the rest of the endoplasm at the observed velocity. Only if the organelles were incorporated in a fibrous or membranous network to improve coupling was a feasible system generated. Yoneda and Nagai (1988) argued that organelles associated with a network of fine, projecting filaments could exert the necessary drag on the endoplasm. Such filaments have been reported after negative staining of endoplasmic material (Nagai and Hayama, 1979) but were not reported in quick-freeze, deep-etched cytoplasm (Kachar and Reese, 1988). Others, however, have invoked the endoplasmic reticulum as the ramifying network that can improve the mechanical coupling between endoplasmic components (Bradley, 1973; Kachar and Reese, 1988; Grolig et al., 1988). (See earlier comments regarding the likely identity of "endoplasmic strands" with endoplasmic reticulum.)

Regulation at the Action Potential

Streaming proceeds at a steady velocity in characean cells except when it is interrupted by the occurrence of an action potential—a propagated depolarization of the electrical potential across both the plasma membrane and, secondarily, the tonoplast. This can be elicited by chemical or physical (mechanical, electrical, or temperature) stimulation. The cytoplasmic streaming abruptly ceases (< 1 s) and recovers over several minutes. A fruitful combination of in vivo and permeabilized cell studies have provided clear evidence that rises in the concentration of cytoplasmic free Ca^{2+} (Ca_c) inhibit streaming. The characean action potential serves as a valuable model of Ca^{2+} regulation in plant cells and will probably emerge after further work as but one example of the influence of Ca^{2+} on the actin-based cytoskeleton in plants.

Studies on Intact Cells

During the action potential, the plasma membrane depolarizes from about 175 mV (inside negative) to near neutrality and repolarizes over a total period of about 2 s (Hope and Walker, 1975). Enhanced Ca^{2+} influx and Cl^- efflux depolarize the membrane, and enhanced K^+ efflux repolarizes the membrane (Hope and Walker, 1975; Tester, 1990). Cytoplasmic streaming ceases at about the time of maximum electrical depolarization of the plasma membrane (Tazawa and Kishimoto, 1968), and since streaming ceases in nonvacuolate cell fragments (Hayama et al., 1979), it is inhibited by consequences flowing from the plasma membrane depolarization rather than from the tonoplast changes. Barry (1969) showed that external Ca^{2+} was required in *Nitella axillaris* for coupling between membrane depolarization and cytoplasmic streaming. (Most species become inexcitable, however, preventing extension of such studies.) The enhanced Ca^{2+} influx at the action potential in *Chara* raises Ca_c from about 0.2 μM to about 7 μM in approximately 400 ms as judged by luminescence from the microinjected photoprotein aequorin (Williamson and Ashley, 1982). Peak Ca_c coincides approximately with maximum electrical depolarization and hence streaming cessation, but there is considerable scope for more refined measurements of the time courses of the three parameters. Rises in Ca_c occur in tonoplast-free cells (Kikuyama and Tazawa, 1983), again consistent with a dependence on the opening of plasma membrane Ca^{2+} channels.

Evidence from both intact cells and membrane-permeabilized and reconstituted motile systems implicates elevated Ca_c as the inhibitor of cytoplasmic streaming. Direct ionophoretic injection of Ca^{2+} into the streaming cytoplasm of *Nitella* locally and reversibly inhibits cytoplasmic streaming (Kikuyama and Tazawa, 1982), and chloroplast rotation is similarly inhibited in injected endoplasmic fragments (Hayama and Tazawa, 1980). A very interesting observation on the time course of motility cessation at the action potential was made by Hayama et al. (1979). Using cells with artificially thickened endoplasm containing many rotating chloroplasts, they reported that the inhibition of chloroplast rotation spreads radially inward with a velocity of 15 $\mu m\ s^{-1}$. It is tempting to think that this represents the speed with which a zone of elevated Ca_c spreads inward from the plasma membrane. There was no such delay in the cessation of streaming at various depths within the thickened layer of endoplasm. This is consistent with the site of action for inhibiting streaming being concentrated around the actin cables where, many years previously, Kamiya and Kuroda (1956) deduced that force was generated in such endoplasm-rich cell fragments. The nature of the changes induced by the action potential in the vicinity of the actin bundles was studied in cells undergoing centrifugation (Kamitsubo et al., 1989). The bulk endoplasm but not the peripheral endoplasm (i.e., that lying nearest to the actin bundles and chloroplasts) is immediately displaced by the centrifugal force when streaming ceases. Preliminary experiments show that this immobilization of the peripheral endoplasm does not occur in regions from which actin cables had earlier been removed by stronger centrifugation. Kamitsubo et al. interpreted the immobilization as requiring the formation of "some kind of transitory cross bridges" between peripheral endoplasm and actin bundles, although this might be better considered as the "freezing" of some normally rapidly cycling cross bridges between an endoplasmic myosin and the actin. Observing only the bulk displacement of endoplasm at the action potential, Tazawa and Kashimoto (1968) had earlier concluded that the motive force disappeared at the action potential, rather than the endoplasm gelating to stop streaming. The argument of Kamitsubo et al. (1989) is that the endoplasm closest to the actin in fact does gelate by binding tightly to the actin cables. It is not clear whether, in addition, the coupling between the bulk and peripheral endoplasm is impaired.

Ca^{2+} Effects in Membrane-Permeabilized Cells

Membrane-permeabilized cells allow experiments to define the ionic conditions that support cytoplasmic streaming. The outlines of the story were established in rapidly perfused cells (Williamson, 1975) in which the large volumes of solution flowing and the extensive extraction of Ca^{2+}-sequestering organelles provide the fullest control of the internal milieu. Reactivation of streaming to in vivo velocities requires approximately neutral pH, about 0.1 μM free Ca^{2+}, millimolar Mg^{2+}, and $\leqslant 80$ mM Cl^-. The conditions have since been studied extensively in both tonoplast- and plasma membrane-permeabilized cells by Shimmen, Tazawa, and colleagues (reviewed in Tazawa and Shimmen, 1987). I will consider only Ca^{2+}, since to date this is the only regulator for which physiologically effective changes in concentration have been documented in living cells.

Ca^{2+} inhibits streaming in all membrane-permeabilized cells, but how they are prepared significantly influences their sensitivity and whether the inhibition is reversed when the Ca^{2+} concentration is subsequently lowered. Rapidly perfused, tonoplast-free cells (see Fig. 3.3a; Williamson, 1975) show 80 percent inhibition of ATP-dependent streaming at 10 μM free Ca^{2+}, but inhibition remains incomplete even at 100 μM Ca^{2+}. Cells are more sensitive to Ca^{2+} when prepared by methods that leave more of the cytoplasm in situ—tonoplast removal by gentle perfusion with slightly hypotonic medium (Tominaga et al., 1987) or plasma membrane permeabilization with a chilled osmoticum containing a Ca^{2+} chelator (see Fig. 3.3c; Tominaga et al., 1983). Inhibition of streaming is complete at 1 μM external Ca^{2+} in cells with a permeabilized plasma membrane (Tominaga et al., 1983). When the concentration of Ca^{2+} is subsequently lowered, rapidly vacuole-perfused cells—

likely to be the most exhaustively extracted—do not resume streaming (Williamson, 1979). In contrast, streaming inhibition in gently perfused and plasma membrane-permeabilized cells is fully reversible (Tominaga et al., 1983, 1987). These differences in response to Ca^{2+} occur in preparations that all sustain streaming at rates comparable to those seen in vivo. It is plausible, but yet to be proved, that certain enzymes and/or Ca^{2+}-binding proteins required for Ca^{2+} sensitivity and reversibility are more readily extracted from permeabilized cells than are the motility-generating cytoskeletal proteins.

Ca^{2+} Effects on Reconstituted Motility

Motility reconstituted by combining exogenous myosins with characean actin cables will be inhibited by Ca^{2+} if the ion has a site of action within the actin cables. If, however, the site at which Ca^{2+} acts in vivo lies on the (missing) characean myosin, the reconstituted motility will show only any sensitivity due to the exogenous myosin. Experimentally, there are no indications that the actin cables have any intrinsic Ca^{2+} sensitivity. Thus, motility reconstituted with characean actin and Ca^{2+}-independent myosins from chick (Vale et al., 1984) or rabbit (Shimmen and Yano, 1985, 1986) skeletal muscles is Ca^{2+} insensitive. The expected Ca^{2+} sensitivity is found if the exogenous myosin is either Ca^{2+} activated (Vale et al., 1984) or Ca^{2+} inhibited (Kohama and Shimmen, 1985), or if the tropomyosin–troponin complex that imparts Ca^{2+} sensitivity to actin filaments in skeletal muscle is bound to the characean actin (Shimmen and Yano, 1985, 1986). Characean actin cables thus support motility at all concentrations of Ca^{2+} at which the exogenous myosin can operate. It therefore appears likely that characean myosin is the target through which Ca^{2+} inhibits cytoplasmic streaming at the action potential, but it has to be conceded that a sensitizing factor could be lost from the actin during the preparative steps preceding bead application. In support of a myosin site of action, elevated Ca^{2+} concentrations in perfused cells alter the distribution of myosin seen by immunofluorescence (Grolig et al., 1988) and remove the filaments associated with the endoplasmic reticulum (Williamson, 1979).

Phosphorylation–Dephosphorylation of Myosin?

Myosins, such as those of mollusks (Szent-Gyorgyi et al., 1973), have intrinsic sensitivity to Ca^{2+}, but Ca^{2+} most commonly acts indirectly via Ca^{2+}-regulated protein kinases that phosphorylate either the light or the heavy chains of myosin IIs (Kuznicki, 1986). Less is known about myosin Is, but the multifunctional Ca^{2+}-binding protein calmodulin is a subunit of at least one member of the myosin I family, the 110,000-M_r mechanoenzyme from brush border (Mooseker and Coleman, 1989).

To date, the only evidence regarding the mode of action of Ca^{2+} on characean myosin comes from physiological experiments. Tominaga et al. (1987) showed that inhibitors of protein phosphatases inhibit streaming in membrane-permeabilized cells even at low Ca^{2+} concentrations and that ATP-γ-S makes irreversible the inhibition due to 10 μM Ca^{2+}. (ATP-γ-S is a nucleotide analog that irreversibly thiophosphorylates proteins because it is a substrate for protein kinases whereas thiophosphorylated proteins are not substrates for protein phosphatases.) Moreover, inhibitors of calmodulin when perfused into cells do not block the inhibition of streaming by elevated Ca^{2+} concentrations, but they do prevent its reversal when the Ca^{2+} concentration is lowered (Tominaga et al., 1985). (Many of the familiar reservations about the lack of specificity of these inhibitors are lessened by their use in permeabilized cells where ATP, Ca^{2+}, and other parameters that might vary in inhibitor experiments in vivo are regulated by the experimenter.) These experiments are consistent with streaming being inhibited by a Ca^{2+}-activated protein kinase that does not depend on calmodulin for its Ca^{2+} activation and being reactivated by a protein phosphatase that is Ca^{2+} activated through calmodulin. A further piece of evidence consistent with this scheme comes from the immunofluorescent localization of calmodulin in whole cells and in cells rapidly vacuole-perfused by the method of Williamson (1975). Calmodulin is widely distributed in the endoplasm of whole cells, although it does not appear to be a component of the actin cables per se (Jablonsky et al., 1990). It is, however, completely undetectable in the rapidly perfused cells in which the reversal of the Ca^{2+} inhibition of streaming (the putative calmodulin-requiring step) is lost (Williamson, 1979).

A potential candidate for the Ca^{2+}-activated, calmodulin-independent protein kinase is the characean homolog of a protein kinase purified and characterized from soybeans (Harmon et al., 1987; Putnam-Evans et al., 1990). Antibodies to this protein bind to fibers in onion and soybean root cells that also stain with fluorescently derivatized phallotoxins (Putnam-Evans et al., 1989), suggesting that it may confer Ca^{2+} sensitivity on the operation of the actomyosin system in higher plants. Harmon and McCurdy (1990) have shown that the same antibodies detect a 50,000-M_r polypeptide in Chara, a similar molecular weight to those recognized in soybeans. The immunoreactive peptide in Chara is more widespread than that in higher plants: actin cables are again strongly labeled, but so is the endoplasm. The pattern is more reminiscent of myosin than of actin but is so widespread as to be nondiagnostic for an association of the two proteins. The putative kinase is, unlike calmodulin, retained in rapidly perfused cells, suggesting a tight binding to some cell component. It remains to confirm the involvement of the kinase in streaming inhibition and to determine its protein substrate(s), but these questions—like so many others—should be accessible through perfused cells.

To postulate a calmodulin-activated phosphatase is at first surprising, since the Ca_c spike at the action potential is quite short-lived compared to the time taken for streaming to recover completely (Williamson and Ashley, 1982). Tominaga et al. (1987) postulated that calmodulin binds Ca^{2+} rapidly but releases it slowly, allowing it to continue activating the postulated phosphatase. The data allow many alternative explanations, such as that the phophatase itself is activated by phosphorylation catalyzed by a calmodulin-activated kinase and only slowly dephosphorylated, perhaps by a Ca^{2+}-independent phosphatase. Alternatively, dephosphorlyation may be rapid and, for example, reassembly of myosin aggregates from dephosphorylated myosin monomer may rate limit the recovery of streaming. It is unlikely that we can progress further until the present data are supplemented by enzymology.

Assembly and Orientation of Actin Bundles

Normal Development

The arrangement of actin in subapical cells that will expand into internodal cells has not been studied by electron microscopy or immunofluorescence but can reasonably be inferred by observing the pattern of streaming. Very short subapical cells already show cytoplasmic streaming proceeding with the same general plan as in the mature internode; that is, streaming is split into upward and downward streams, each occupying 180° of the cell's circumference. Chloroplasts take up a cortical position at an early stage (Pickett-Heaps, 1975), but at this stage there is not a distinct gap in the chloroplast arrays forming a well-defined neutral line. The origin of that initial arrangement of both chloroplasts and actin cables is unknown. Cells twist as they elongate, leading Green (1954, 1963, 1964) to propose that mechanical strain occurring in growing cells generates and aligns the helical features of the cortical cytoplasm—that is, individual chloroplasts, chloroplast files, neutral line, and the (at that time) unknown structures generating cytoplasmic streaming. A purely physical process thus aligns them parallel to the cell's main axis of elongation. (The realignment of cellulose microfibrils from transverse to longitudinal in growing cells may be a more familiar example of the same effect.) The effects of the directionality of cell growth were most directly demonstrated in the case of chloroplasts that elongate parallel to the files in normally growing cells but increase in width in cells where transverse growth exceeds longitudinal growth (Green, 1964). Microtubules are strongly implicated in constraining cells to grow as elongate cylinders (Green et al., 1970; Wasteneys, 1988), but it is unknown why the cell twists as it grows, so generating the helical arrangement of chloroplast files, actin cables, and neutral lines.

Bundle Assembly in Maturing Internodal Cells

The number of actin bundles per chloroplast file remains nearly constant as the length of *Chara* internodal cells increases from 1 to 5 cm (Williamson and Hurley, 1986). As cell growth often (Green, 1954) but not always (Taiz et al., 1981) occurs uniformly throughout the length of the cell, it seems likely that actin bundles will similarly add new filaments along their length as they are stretched by cell growth. A first attempt to study this process in expanding internodal cells of *Chara* was made by using cytochalasin B (Williamson & Hurley, 1986), a fungal metabolite that binds to the barbed ends of actin filaments that are capable of rapid extension by monomer addition (MacLean-Fletcher and Pollard, 1980; Pollard and Mooseker, 1981). In long-term experiments, cytochalasin reduces the number of actin bundles beneath each file of chloroplasts from 5.2 to 2.0 in growing cells and to 3.4 in nongrowing cells. It was notable, however, that cytochalasin-affected bundles did not fragment (as often seen in animal cells) and the lengths of each bundle increased during cell extension in spite of the presence of cytochalasin. It was concluded that bundle growth in extending cells does not have an obligate requirement for the extension of barbed filament ends, a process that cytochalasin blocks. The effects of cytochalasin B on characean cells are probably complex and are discussed by Williamson and Hurley (1986).

Assembly of Regenerating Bundles

Rather different questions can be asked in large internodal cells, where the actin cables can be locally disrupted by relatively simple techniques and their reassembly can be studied. Centrifugation was used before the discovery of the actin cables to exfoliate the cortical chloroplasts and locally inhibit streaming and contributed to the successful recognition of actin cables in living cells with phase contrast microscopy (Kamitsubo, 1966). Strong, local irradiation with the microscope lamp provides a method to control the position and size of the area from which chloroplasts and actin bundles are removed (Kamitsubo, 1972b). In relatively small "windows" (i.e., optically favorable areas lacking chloroplasts) produced by this technique, actin bundles regenerate over about 24 h, but no details of the regeneration process were deduced. When larger "windows" are produced and regeneration followed by immunofluorescence is carried out at varying times after wounding

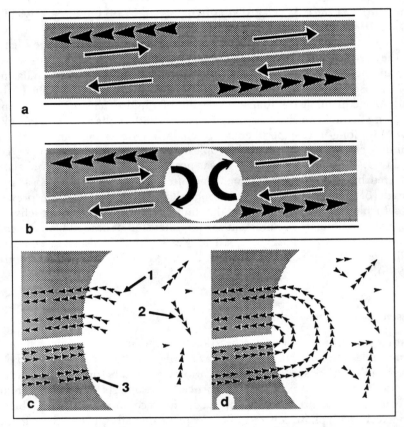

(Fig. 3.4; Williamson et al., 1984), the observations suggest two origins for regenerating bundles: preexisting bundles severed at the edge of the wound extend into the wound, and new bundles are initiated in the "window" that are not connected to preexisting bundles. Studying wounds spanning the neutral line shows that only bundles that contain actin filaments with exposed plus ends extend into the window; only a few micrometers away across the neutral line, bundles in which minus ends are exposed do not

Figure 3.4. Actin bundle polarity, endoplasmic flow and bundle regeneration in wounded areas of *Chara*. (a) The original streaming pattern (arrows) and bundle polarities (arrowhead filaments). (b) After local bundle destruction, endoplasm follows curving paths but is stationary in the central regions. (c) Filament bundles with exposed barbed filament ends extend (1) by a cytochalasin-sensitive mechanism while those exposing pointed filament ends (3) show little growth. Assembly of bundles unconnected with pre-existing bundles (2) is less sensitive to cytochalasin. (d) Actin bundles at the edge of the wounded region grow in smooth curves that reflect and augment the curving flow paths seen in (b) while those in the non-flowing central regions are disorganised.

extend. Extension of the bundles with exposed plus ends is completely inhibited by concentrations of cytochalasin B that are insufficient to inhibit streaming (Williamson and Hurley, 1986), but short lengths of actin bundle are scattered through the wounded area (see Fig. 3.4c).

These findings are consistent with the following:

1. Bundle regeneration occurring by extension of the bundles whose filaments have barbed ends exposed where they are severed at the edge of the wound.
2. Cytochalasin being able to block this process, probably by acting as a capping agent for barbed filament ends as documented for animal actin in vitro (MacLean-Fletcher and Pollard, 1980; Pollard and Mooseker, 1981).
3. Bundles that are not continuous with preexisting bundles also being initiated in the wound. Such bundles grow by a process that is not completely blocked by cytochalasin and so is probably not obligatorily dependent on the extension of the barbed filament ends.
4. The bundles that have the pointed ends of their actin filaments exposed possibly being prevented from extending either by capping proteins or by the concentration of actin being too low to extend that low affinity end of the filament.

Although this begins to define the bundle assembly process, a considerable range of options for bundle assembly mechanisms still exists (e.g., filament elongation could occur before and/or after filaments are cross-linked into the bundle and could involve either monomer addition or the annealing of short filaments into longer ones).

Alignment of Regenerated Actin Bundles

The alignment of actin bundles following cell damage inflicted by centrifugation or intense illumination depends on the size of the disrupted area. Small "windows" across which cytoplasm was previously flowing in one direction only (Kamitsubo, 1972b; Allen, 1974) regenerate actin bundles that run straight across the window and restore the continuity of streaming. At the other extreme, a cell centrifuged to remove the great bulk of its chloroplasts regenerates a disordered pattern of streaming that shows no overall organization at the whole cell level (Chen, 1983). Different regions of the cell can show streaming at any angle relative to the original axis, and circular patterns are set up in places. Interestingly, the chloroplasts subsequently align along the abnormal pathways for streaming.

Studies of windows larger than those studied previously by Kamitsubo and by Allen and positioned to span the neutral line led to hypotheses for the alignment of regenerating bundles that have relevance to normal morphogenesis (Williamson et al., 1984). Instead of reestablishing the original straight actin bundles, the regenerating actin frequently regrows in a curving pattern in these large windows, so that it contacts the actin on the other side of the neutral line. This alignment parallels the flow of endoplasm present in the newly made window (see Fig. 3.4b), suggesting that the actin regrowing from the severed bundle ends is aligned by flow generated by the surviving actin. The actin that regenerates in the center of the window is in a region in which there is no such flow at early stages and it does not show a dominant alignment. It is therefore postulated that actin is self-aligning as a result of the flows that it itself generates. In this context, the significance of the regeneration from bundles exposing plus ends but not those exposing minus ends becomes apparent. Because of the relationship between actin polarity and streaming direction (Kersey et al., 1976), growth of bundles that have exposed minus ends would be upstream, against the flow of the cytoplasm and therefore prone to be forced into a U turn if the growing bundles deviated even slightly from parallelism with the flow. Bundles in which plus ends are exposed will extend in the same direction as the flow and therefore be subject to forces restoring them to parallelism if they deviate from the direction of flow.

This proposal for flow-induced self-alignment does not supplant the strain alignment proposed by Green but may nonetheless be relevant to the ontogeny of the normal cell. First, self-induced flow alignment could restrict the tendency of actin cables to arise at aberrant angles with respect to the bulk of the actin that is undergoing strain alignment and supporting rapid flow of endoplasm. Second, the end walls of an internodal cell will undergo only limited and virtually isotropic strain, so providing no strong orienting mechanism for the actin that crosses the end walls. Actin on the longitudinal walls, by delivering and removing endoplasm from the end walls, would be expected to establish a U-turn flow across the end walls that could, by a process exactly analogous to that seen in the large windows, align actin growing across that end wall.

In summary, no studies of actin regeneration support the view that there is a template directing actin assembly and orientation in the cortex, and the neutral line does not have any special status in terms of inhibiting bundle regeneration in mature cells. Rather, it appears that actin has a tendency to self-assemble into bundles (presumably by reaction with a number of so far unknown proteins), that they can be anchored to the cortex in the presence or absence of chloroplasts, and that strain and flow-induced self-alignment operate to orient them. When regeneration occurs too rapidly and/or in cells that have ceased growing, self-induced flow alignment dominates. This regenerates the original streaming pattern in small windows but readily forms aberrant patterns in larger windows and when the entire cortical array is dislodged by centrifugation. This ability of actin to assemble in aberrant patterns contrasts with the ability of microtubules to return accurately to a transverse orientation when they reassemble after depolymerization (Wasteneys and Williamson, 1989).

Conclusions

Characean algae have, with occasional exceptions, proved resistant to preparative biochemistry and have not been exploited for genetics or molecular biology. Nonetheless, because they have been successfully adapted to the changing demands of experimenters, they continue to be the plant cells in which cytoplasmic streaming is most fully understood. Increasingly, progress depends on a synergism between work on other organisms and on the Characeae. Antibodies to proteins purified and characterized from other, biochemically more convenient sources have played an important role in elucidating characean streaming, but, equally, defining the function of those proteins is often easier in the Characeae than in the organism from which they were originally purified. There are few reasons to believe that this synergistic partnership will not continue to illuminate streaming in the Characeae and provide a source of ideas for studies on motility in the rest of the plant kingdom.

Acknowledgments
I thank James Whitehead for help with the figures.

References

Adams, R. J., and Pollard, T. D. 1986. Propulsion of organelles isolated from *Acanthamoeba* along actin filaments by myosin–1. *Nature (Lond.)* 322: 754–756.

Adams, R. J. and Pollard, T. D. 1989. Binding of myosin I to membrane lipids. *Nature (Lond.)* 340: 565–568.

Albanesi, J. P., Fujisaki, H., Hammer III, J. A., Korn, E. D., Jones, R., and Sheetz, M. P. 1985. Monomeric *Acanthamoeba* myosins I support movement *in vitro*. *J. Biol. Chem.* 260: 8649–8652.

Allen, N. S. 1984. Endoplasmic filaments generate the motive force for rotational streaming in *Nitella*. *J. Cell Biol.* 63: 270–287.

Barry, W. H. 1969. Coupling of excitation and cessation of cyclosis in *Nitella:* Role of divalent cations. *J. Cell Physiol.* 72: 153–160.

Bradley, M. O. 1973. Microfilaments and cytoplasmic streaming: inhibition of streaming by cytochalasin. *J. Cell. Sci.* 12: 327–343.

Chen, J. C. W. 1983. Effect of elevated centrifugal field on the *Nitella* cell and postcentrifugation patterns of its cytoplasmic streaming and chloroplast files. *Cell Struct. Funct.* 8: 109–118.

Chen, J. C. W., and Kamiya, N. 1975. Localization of myosin in the internodal cell of *Nitella* as suggested by differential treatment with N-ethylmaleimide. *Cell Struct. Funct.* 1: 1–9.

Dabora, S. L., and Sheetz, M. P. 1988. The microtubule-dependent formation of a tubulovesicular network with characteristics of the ER from cultured cell extracts. *Cell* 54: 27–35.

Green, P. B. 1954. The spiral growth pattern of the cell wall in *Nitella axillaris. Am J. Bot.* 41: 403–409.

Green, P. B. 1963. On mechanisms of cellular elongation. In M. Locke, ed., *Cytodifferentiation and Macromolecular Synthesis.* Academic Press, New York, pp. 203–234.

Green, P. B. 1964. Cinematic observations on the growth and division of chloroplasts in *Nitella. Am. J. Bot.* 51: 334–342.

Green, P. B., Erickson, R. O., and Richmond, P. A. 1970. On the physical basis of cell morphogenesis. *Ann. N.Y. Acad. Sci.* 175: 712–731.

Grolig, F., Williamson, R. E., Parke, J., Miller, C., and Anderton, B. H. 1988. Myosin and Ca^{2+}-sensitive streaming in the alga *Chara:* two polypeptides reacting with a monoclonal anti-myosin and their localization in the streaming endoplasm. *Eur. J. Cell Biol.* 47: 22–31.

Harmon, A. C., and McCurdy, D. W. (1990). Calcium-dependent protein kinase and its possible role in the regulation of the cytoskeleton. *Current Topics in Plant Biochemistry and Physiology* 9: 119–128.

Harmon, A. C., Putnam-Evans, C., and Cormier, M. J. 1987. A calcium-dependent but calmodulin-independent protein kinase from soybean. *Plant Physiol. (Bethesda)* 83: 830–837.

Hayama, T., and Tazawa, M. (1980). Ca^{2+} reversibly inhibits active rotation of chloroplasts in isolated cytoplasmic droplets of *Chara. Protoplasma* 102: 1–9.

Hayama, T., Shimmen, T., and Tazawa, M. 1979. Participation of Ca^{2+} in cessation of cytoplasmic streaming induced by membrane excitation in Characeae internodal cells. *Protoplasma* 99: 305–321.

Hayashi, T. 1964. The role of the cortical gel layer in cytoplasmic streaming. In R. D. Allen and N. Kamiya, ed., *Primitive Motile Systems in Cell Biology.* Academic Press, New York, pp. 19–29.

Higashi-Fujime, S. 1980. Active movement *in vitro* of microfilaments isolated from *Nitella* cell. *J. Cell Biol.* 87: 569–578.

Hope, A. B., and Walker, N. A. (1975). *The Physiology of Giant Algal Cells.* Cambridge University Press, London, 201 pp.

Huxley, H. E. (1963). Electron microscope studies on the structure of natural and synthetic protein filaments from striated muscle. *J. Mol. Biol.* 7: 281–308.

Jablonsky, P. P., Hagan, R. P., Grolig, F., and Williamson, R. E. 1990. Immunolocalization of *Chara* calmodulin and the reversibility of the inhibition of cytoplasmic streaming by Ca^{2+}. In *Calcium in Plant Growth and Development,* 13th Annual Symposium in Plant Physiology, American Society of Plant Physiology, pp. 79–85.

Jarosch, R. 1956. Plasmastromung und Chloroplastenrotation bei Characeen. *Phyton (B. Aires)* 8: 87–107.

Kachar, B. 1985. Direct visualization of organelle movement along actin filaments dissociated from characean algae. *Science* 227: 1355–1357.

Kachar, B., and Reese, T. S. 1988. The mechanism of cytoplasmic streaming in characean algal cells: Sliding of endoplasmic reticulum along actin filaments. *J. Cell Biol.* 106: 1545–1552.

Kamitsubo, E. 1966. Motile protoplasmic fibrils in cells of *Characeae*. II. Linear fibrillar structure and its bearing on protoplasmic streaming. *Proc. Jpn. Acad.* 42: 640–643.

Kamitsubo, E. 1972a. Motile protoplasmic fibrils in cells of the Characeae. *Protoplasma* 74: 53–70.

Kamitsubo, E. 1972b. A "window technique" for detailed observation of cytoplasmic streaming. *Exp. Cell Res.* 74: 613–616.

Kamitsubo, E., Ohashi, Y. & Kikuyama, M. 1989. Cytoplasmic streaming in internodal cells of *Nitella* under centrifugal acceleration: A study done with a newly constructed centrifuge microscope. *Protoplasma* 152: 148–155.

Kamiya, N. 1959. Protoplasmic streaming. In L. V. Heilbrunn and F. Weber (ed.), *Protoplasmatologia,* Vol. VIII/3a. Springer-Verlag, Vienna, 199 pp.

Kamiya, N. 1960. Physics and chemistry of protoplasmic streaming. *Annu. Rev. Plant Physiol.* 11: 323–340.

Kamiya, N. 1962. Protoplasmic streaming. In W. Ruhland (ed.), *Encyclopaedia of Plant Physiology*. Vol. 17/2. Springer-Verlag, Berlin, pp. 979–1035.

Kamiya, N. 1981. The physical and chemical basis of cytoplasmic streaming. *Annu. Rev. Plant Physiol.* 32: 205–236.

Kamiya, N. 1986. Cytoplasmic streaming in giant algal cells: A historical survey of experimental approaches. *Bot. Mag. Tokyo* 99: 441–467.

Kamiya, N., and Kuroda, K. 1956. Velocity distribution of the protoplasmic streaming in *Nitella* cells. *Bot. Mag. Tokyo* 69: 544–554.

Kamiya, N., and Kuroda, K. 1958. Measurement of the motive force of the protoplasmic rotation in *Nitella*. *Protoplasma* 50: 144–148.

Kamiya, R., and Nagai, R. 1982. Structural similarity between actin bundles from characean algal cells and sea urchin oocytes. *J. Mol. Biol.* 155: 169–172.

Kamiya, N., and Tazawa, M. 1966. Surgical operations on characean cells with special reference to cytoplasmic streaming. *Annual Report Faculty of Science, Osaka University* 14: 1–37.

Kato, T., and Tonomura, Y. 1977. Identification of myosin in *Nitella flexilis*. *J. Biochem.* (Tokyo) 82: 777–782.

Kersey, Y. M., Hepler, P. K., Palevitz, B. A., & Wessels, N. K. 1976. Polarity of actin filaments in characean algae. *Proc. Natl. Acad. Sci. USA* 73: 165–167.

Kikuyama, M., and Tazawa, M. 1982. Ca^{2+} ion reversibly inhibits the cytoplasmic streaming of *Nitella*. *Protoplasma* 113: 241–243.

Kikuyama, M., and Tazawa, M. 1983. Transient increase of intracellular Ca^{2+} during excitation of tonoplast-free *Chara* cells. *Protoplasma* 117: 62–67.

Kohama, K., and Shimmen, T. 1985. Inhibitory Ca^{2+}-control of movement of beads coated with *Physarum* myosin along actin cables in *Chara* internodal cells. *Protoplasma* 129: 88–91.

Kohno, T., and Shimmen, T. 1988. Accelerated sliding of pollen tube organelles along Characeae actin bundles regulated by Ca^{2+}. *J. Cell Biol.* 106: 1539–1543.

Korn, E. D., and Hammer III, J. A. 1988. Myosins of non-muscle cells. *Ann. Rev. Biophys. Biophys. Chem.* 17: 23–45.

Kron, S. J., and Spudich, J. A. 1986. Fluorescent actin filaments move on myosin fixed to a glass surface. *Proc. Natl. Acad. Sci. USA* 83: 6272–6276.

Kuroda, K. 1983. Cytoplasmic streaming in characean cells cut open by microsurgery. *Proc. Jpn. Acad. Ser. B* 59: 126–130.

Kuroda, K., and Kamiya, N. 1975. Active movement of *Nitella* chloroplast *in vitro*. *Proc. Jpn. Acad. Ser. B* 51: 774–777.

Kuroda, K., and Kamiya, N. 1981. Behaviour of cytoplasmic streaming in *Nitella* during centrifugation as revealed by the television centrifuge-microscope. *Biorheology* 18: 633–641.

Kuznicki, J. 1986. Phosphorylation of myosin in non-muscle and smooth muscle cells. *FEBS Lett.* 204: 169–176.

MacLean-Fletcher, S., and Pollard, T. 1980. Mechanism of action of cytochalasin B on actin. *Cell* 20: 329–341.

McLean, B., and Juniper, B. E. 1988. Fine structure of *Chara* actin bundles, using rapid freezing and deep etching. *Cell Biol. Int. Rep.* 12: 509–517.

Mooseker, M. S., and Coleman, T. R. 1989. The 110-kD protein-calmodulin complex of the intestinal microvillus (brush border myosin I) is a mechanoenzyme. *J. Cell Biol.* 108: 2395–2400.

Nagai, R., and Hayama, T. 1979. Ultrastructure of the endoplasmic factor responsible for cytoplasmic streaming in *Chara* internodal cells. *J. Cell Sci.* 36: 121–136.

Nagai, R, and Kamiya, N. 1977. Differential treatment of *Chara* cells with cytochalasin B with special reference to its effects on cytoplasmic streaming. *Exp. Cell Res.* 108: 231–237.

Nagai, R., and Rebhun, L. I. 1966. Cytoplasmic microfilaments in streaming *Nitella* cells. *J. Ultrastruct. Res.* 14: 571–589.

Nothnagel, E. A., Barak, L. S., Sanger, J. W., and Webb, W. W. 1981. Fluorescence studies on modes of cytochalasin B and phallotoxin action on cytoplasmic streaming in *Chara*. *J. Cell Biol.* 88: 364–372.

Nothnagel, E. A., and Webb, W. W. 1982. Hydrodynamic models of viscous coupling between motile myosin and endoplasm in characean algae. *J. Cell Biol.* 94: 444–454.

Owaribe, K., Izutsu, K., and Hatano, S. 1979. Cross-reactivity of antibody to *Physarum* actin and actins in eukaryotic cells examined by immunofluorescence. *Cell Struct. Funct.* 4: 117–126.

Palevitz, B. A., Ash, J. F., and Hepler, P. K. 1974. Actin in the green algae, *Nitella Proc. Natl. Acad. Sci. U.S.A.* 71: 363–366.

Palevitz, B. A., and Hepler, P. K. 1975. Identification of actin *in situ* at the ectoplasm-endoplasm interface of *Nitella*. Microfilament–chloroplast association. *J. Cell Biol.* 65: 29–38.

Pickett-Heaps, J. D. 1967. Ultrastructure in *Chara* sp. I. Vegetative cells. *Aust. J. Biol. Sci.* 20: 539–551.

Pickett-Heaps, J. D. 1975. *Green Algae*. Sinauer, Sunderland MA, 606 pp.

Pollard, T. D., and Cooper, J. A. (1986). Actin and actin-binding proteins. A critical evaluation of mechanisms and functions. *Annu. Rev. Biochem.* 55: 987–1035.

Pollard, T. D., and Mooseker, M. S. 1981. Direct measurements of actin polymerization rate constants by electron microscopy. Filaments nucleated by isolated microvillus cores. *J. Cell Biol.* 88: 654–659.

Putnam-Evans, C., Harmon, A. C., and Cormier, M. J. 1990. Purification and characterization of a novel protein kinase from soybean. *Biochemistry* 29: 2488–2495.

Putnam-Evans, C., Harmon, A. C., Palevitz, B. A., Fechheimer, M., and Cormier, M. J. 1989. Calcium-dependent protein kinase is localized with F-actin in plant cells. *Cell Motil. Cytoskel.* 12: 12–22.

Sakano, K., and Tazawa, M. 1986. Tonoplast origin of the envelope membrane of cytoplasmic droplets prepared from *Chara* internodal cells. *Protoplasma* 131: 247–249.

Sheetz, M. P., and Spudich, J. A. 1983. Movement of myosin-coated fluorescent beads on actin cables *in vitro*. *Nature (Lond.)* 303: 31–35.

Shimmen, T. 1988. Characean actin bundles as a tool for studying actomyosin-based motility. *Bot. Mag. Tokyo* 101: 533–544.

Shimmen, T., and Tazawa, M. 1982. Reconstitution of cytoplasmic streaming in Characeae. *Protoplasma* 113: 127–131.

Shimmen, T., and Yano, M. 1984. Active sliding movement of latex beads coated with skeletal muscle myosin on *Chara* actin bundles. *Protoplasma* 121: 132–137.

Shimmen, T., and Yano, M. 1985. Ca^{2+} regulation of myosin sliding along *Chara* actin mediated by native tropomyosin. *Proc. Jpn. Acad. Ser. B* 61: 86–89.

Shimmen, T., and Yano, M. 1986. Regulation of myosin sliding along *Chara* actin bundles by native skeletal muscle tropomyosin. *Protoplasma* 132: 129–136.

Szent-Gyorgyi, A. G., Szentkiralyi, E. M., and Kendrick-Jones, J. 1973. The light chains of scallop myosin as regulatory subunits. *J. Mol. Biol.* 74: 179–203.

Taiz, L., Metraux, J.-P., and Richmond, P. A. 1981. Control of cell expansion in the *Nitella* internode. In O. Kiermayer (ed.), *Cytomorphogenesis in plants*. Springer-Verlag, Vienna, pp. 231–64.

Tazawa, M. 1964. Studies on *Nitella* having artificial cell sap. I. Replacement of the cell sap with artificial solutions. *Plant Cell Physiol.* 5: 33–43.

Tazawa, M. 1968. Motive force of the cytoplasmic streaming in *Nitella*. *Protoplasma* 65: 207–222.

Tazawa, M., Kikuyama, M., and Shimmen, T. 1976. Electric characteristics and cytoplasmic streaming of Characeae cells lacking tonoplast. *Cell Struct. Funct.* 1: 165–176.

Tazawa, M., and Kishimoto, U. 1968. Cessation of cytoplasmic streaming of *Chara* internodes during action potential. *Plant Cell Physiol.* 9: 361–368.

Tazawa, M., and Shimmen, T. 1987. Cell motility and ionic relations in characean cells as revealed by internal perfusion and cell models. *Int. Rev. Cytol.* 109: 259–312.

Tester, M. 1990. Plant ion channels: Whole-cell and single-channel studies. *New Phytol.* 114: 305–340.

Titus, M. A., Warrick, H. M., and Spudich, J. A. 1989. Multiple actin-based motor genes in *Dictyostelium*. *Cell Regul.* 1: 55–63.

Tominaga, Y., Muto, S., Shimmen, T., and Tazawa, M. 1985. Calmodulin and Ca^{2+}-controlled cytoplasmic streaming in characean cells. *Cell Struct. Funct.* 10: 315–325.

Tominaga, Y., Shimmen, T., and Tazawa, M. 1983. Control of cytoplasmic streaming by extracellular Ca^{2+} in permeabilized *Nitella* cells. *Protoplasma* 116: 75–77.

Tominaga, Y., Wayne, R., Tung, H. Y. L., and Tazawa, M. 1987. Phosphorylation-dephosphorylation is involved in Ca^{2+}-controlled cytoplasmic streaming of characean cells. *Protoplasma* 136: 161–169.

Vale, R. D., Szent-Gyorgyi, A. G., and Sheetz, M. P. 1984. Movement of scallop myosin on *Nitella* actin filaments: Regulation by calcium. *Proc. Natl. Acad. Sci. U.S.A.* 81: 6775–6778.

Wasteneys, G. O. 1988. Microtubule organization in internodal cells of characean algae. Ph.D. Thesis, Australian National University, Canberra, 198 pp.

Wasteneys, G. O., and Williamson, R. E. (1989). Reassembly of microtubules in *Nitella tasmanica*: Quantitative analysis of assembly and orientation. *Eur. J. Cell Biol.* 50: 76–83.

Wessels, N. K., Spooner, B. S., Ash, J. F., Bradley, M. O., Luduena, M. A., Taylor, E. L., Wrenn, J. T., and Yamada, K. M. 1971. Microfilaments in cellular and developmental processes. *Science* 171: 135–143.

98   Williamson

Williamson, R. E. 1972. A light microscope study of the action of cytochalasin B on the cells and isolated cytoplasm of the Characeae. *J. Cell Sci.* 10: 811–819.

Williamson, R. E. 1974. Actin in the alga, *Chara corallina*. *Nature (Lond.)* 248: 801–802.

Williamson, R. E. 1975. Cytoplasmic streaming in *Chara:* A cell model activated by ATP and inhibited by cytochalasin B. *J. Cell Sci.* 17: 655–668.

Williamson, R. E. 1979. Filaments associated with the endoplasmic reticulum in the streaming cytoplasm of *Chara corallina*. *Eur. J. Cell Biol.* 20: 177–183.

Williamson, R. E. 1985. Immobilisation of organelles and actin bundles in the cortical cytoplasm of the alga *Chara corallina* Klein ex. Wild. *Planta (Berl.)* 163: 1–8.

Williamson, R. E., and Ashley, C. C. 1982. Free Ca^{2+} and cytoplasmic streaming in the alga *Chara*. *Nature (Lond.)* 296: 647–651.

Williamson, R. E., and Hurley, U. A. 1986. Growth and regrowth of actin bundles in *Chara:* bundle assembly by mechanisms differing in sensitivity to cytochalasin B. *J. Cell Sci.* 85: 21–32.

Williamson, R. E., Hurley, U. A., and Perkin, J. L. 1986. Regeneration of actin bundles in *Chara:* polarized growth and orientation by endoplasmic flow. *Eur. J. Cell Biol.* 34: 221–228.

Williamson, R. E., McCurdy, D. W., Hurley, U. A., and Perkin, J. L. 1987. Actin of *Chara* giant internodal cells. *Plant Physiol. (Bethesda)* 85: 268–72.

Williamson, R. E., Perkin, J. L., McCurdy, D. W., Craig, S., and Hurley, U. A. 1986. Production and use of monoclonal antibodies to study the cytoskeleton and other components of the cortical cytoplasm of *Chara*. *Eur. J. Cell Biol.* 41: 1–8.

Williamson, R. E., and Toh, B. H. 1979. Motile models of plant cells and the immunofluorescent localization of actin in a motile *Chara* cell model. In S. Hatano, H. Ishikawa, and H. Sato (ed.), *Cell Motility: Molecules and Organization.* University of Tokyo Press, Tokyo, pp. 339–46.

Yoneda, M., and Nagai, R. 1988. Structural basis of cytoplasmic streaming in characean internodal cells. A hydrodynamic analysis. *Protoplasma* 147: 64–76.

Chapter 4

Flagellar Beat Patterns in Algae

Stuart F. Goldstein

Introduction

During the last several decades there has been an explosive increase in knowledge of the ultrastructural details of algal cells. There has also been an impressive advance in knowledge of the ultrastructure of flagella and of the mechanism of flagellar beating. However, there has not been a comparable advance in knowledge of the waveforms of flagellar beating in algae, despite the existence of the required techniques. The major exception is the work that has been done on *Chlamydomonas,* which has been used as a model system in studies of flagellar motility. In this review, no attempt is made to catalogue completely the many incidental descriptions of flagellar waveforms that have been reported in the course of ultrastructural and taxonomic studies. But in many cases they are the only reported descriptions, and they are used here where they represent the published information on particular groups.

The basic algal configuration consists of two flagella emerging next to one another from the cell body. The flagella can be *equal* (the same length) or *unequal* (of different length). They can be *homodynamic* (beating with the same waveform) or *heterodynamic* (beating with different waveforms). The flagellar surface can contain a variety of structures, including *mastigonemes* (rigid tubular projections), *hairs* (flexible nontubular projections), and scales. The mastigonemes, hairs, and combinations of flagella with similar or dissimilar surfaces have been given a variety of names; these names have been reviewed by Moestrup (1982).

Algal flagella generally beat too rapidly for their waveforms to be discerned with continuous illumination. Observation is greatly aided by stroboscopic illumination, although recording techniques with good temporal resolution are required for quantitative analyses (e.g., Goldstein, 1983). They often exhibit rapid changes in motility. High-speed recording techniques are necessary for detailed analyses of these rapid transient events.

Descriptions

Chrysophyceae

The chrysophytes commonly have an anterior flagellum that bears two opposing rows of mastigonemes, plus a second, smooth flagellum (Leedale et al., 1970). The mastigoneme-bearing flagellum is usually longer than the smooth one, projects anteriorly, and is usually the flagellum that propels the cell. The smooth flagellum is typically inserted next to the other one and is roughly perpendicular to it and curves back toward the cell body; it is vestigal or absent in some species.

The structure of the flagella of *Ochromonas* has been studied in some detail. Like a number of other chrysomonads, it has a longer, mastigoneme-bearing flagellum and a shorter, smooth flagellum (Manton, 1952; Pitelka and Schooley, 1955; Bouck, 1971). The mastigonemes lie in two rows, one on either side of the longer flagellum. The longer flagellum is roughly 15 μm long; the shorter one is about 5 μm long.

Ochromonas swims with the longer flagellum held forward; that is, the flagellum pulls the cell (Jahn and Bovee, 1964; Jahn et al., 1964b; Holwill and Sleigh, 1967, Bouck, 1971). When the cell is attached to a substrate, water and food particles near the long flagellum are moved toward the cell. The shorter flagellum lies along the anterior side of the cell and is motionless during swimming. O. *minima* swims with a speed of 75 μm s^{-1} and gyrates 0.25–1.25 times per second (Throndsen, 1973).

The motility of chrysomonad flagella has been studied visually under stroboscopic illumination and with photographic film. Jahn et al. (1964b), studying O. *malhamensis*, O. *danica*, and an unidentified chrysomonad thought to be *Chromulina*, found that waves on the long flagellum are planar and pass from the basal (cell body) end toward the tip. Using high-speed photography at up to 600 frames per second, they found the long flagellum of O. *malhamensis*, O. *danica*, and *Chromulina* (?) to beat at 50 Hz, 50 Hz, and 27 Hz, respectively. They described the waveforms as sine waves. Their tracings show the waveforms to be somewhat irregular, in contrast to the regular waveforms shown in the photographs of O. *danica* by Bouck (1971); this may be due to the intense illumination needed for the high-speed films.

Holwill and Sleigh (1967) have seen planar waves passing from base to tip in *O. malhamensis* and an unidentified chrysomonad. They found the long flagellum of *O. malhamensis* and the unidentified chrysomonad to beat at 68.4 Hz (18°C) and 57.0 Hz (20°C), respectively. They did not show photographs or tracings but gave detailed waveform measurements. In both organisms the amplitude and wavelength increased as the waves moved from base to tip. The amplitude increased from 0.65 μm to 1.25 μm in *O. malhamensis* and from 0.75 μm to 1.15 μm in the other chrysomonad; the wavelength increased from 4.0 μm to 9.0 μm in *O. malhamensis* and from 3.2 μm to 8.9 μm in the other chrysomonad. Holwill and Peters (1974) found the frequency, amplitude, and propulsive speed—but not the wavelength—of *O. danica* to decrease as the viscosity of the medium increased.

Herth (1982) reported that in the distally propagating wave of the long flagellum of *O. malhamensis*, "the curved region of the flagellum describes a circular movement. This is a complex three dimensional beat, with a strong uniplanar component." Observations were aided by flash photographs, but not by stroboscopic illumination or high-speed recordings. The reason for the difference between this and previous reports is not clear. The short flagellum was found to describe "only a small forward–backward or circular movement with no obvious sinuous bending."

The exact waveforms of *O. danica* (Fig. 4.1a) have not been characterized. Differential interference flash photographs of the long flagellum of live cells by Bouck (1971) give the impression that its waveform would be better approximated by a series of circular arcs and interconnecting straight regions than by sine waves (cf. Brokaw and Wright, 1963; see the following discussion of the Dinophyceae).

Passage of planar waveforms from base to tip on the long flagellum has also been seen in *Actinomonas* (Fig. 4.1b) and *Poteriodendron* (Fig. 4.1c) (both of which bear a single flagellum) and in *Monas* (Fig. 4.1d), by Sleigh (1964). He found their beat frequencies to be about 50 Hz (18°C), 50 Hz (20°C), and 40–50 Hz, respectively. He gave sketches and approximate measurements of their waveforms. Their lengths were about 20–30 μm, 25–

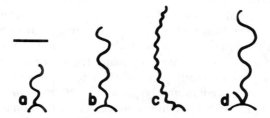

Figure 4.1. Waveforms of some chrysophytes. (a) *Ochromonas danica*. (b) *Actimonas*. (c) *Poteriodendron*. (d) *Monas*. Bar = 10 μm. (a after Bouck, 1971; b, c, and d after Sleigh, 1964.)

30 μm, and 25–40 μm, respectively. The wavelength of *Actinomonas* was about 7–10 μm. The wavelength of *Monas* was about 6 μm at the base and increased to 10–12 μm distally. He did not confirm the earlier assertion by Lowndes (1944) that the waveform and swim path are spiral; he found the waveform to be planar, like those of other chrysomonads. The flagellum of *Poteriodendron* was slightly curved and had small-amplitude waves; the wavelength was 3–5 μm and tended to decrease distally. The cell body occupied a lorica and was attached by a contracile foot thought to be a modified flagellum. When the cell was stimulated, it retracted into the lorica and the motile flagellum rapidly coiled into a plane spiral; upon relaxation the flagellum slowly uncoiled over a period of a second or two.

Wetherbee et al. (1988) described the long, hairy flagellum of *Epipyxis pulchra* as exhibiting a rapid sigmoid waveform, shown as a single S shape along the entire flagellum in the photograph of their Fig. 2. In several of their photographs of live cells the long flagellum has roughly sinusoidal waves, and is about one and a half wavelengths long. The short, smooth flagellum had an irregular beat, but it occasionally exhibited a sigmoid beat similar to the sigmoid form of the long flagellum.

In all these organisms, if the cell is free-swimming it is pulled in the direction of movement of the flagellar waves; if the cell is attached to a substrate, water and food particles in the vicinity of the flagellum are moved in the opposite direction, toward the cell body. This is a somewhat surprising result; in "typical" flagella, such as those of invertebrate spermatozoa (e.g., Gray, 1955), the water is moved in the direction of the traveling wave (toward the tip) and the cell is propelled in the opposite direction. As pointed out by Jahn et al. (1964b), the reversal of propulsive forces in these flagella is due to their mastigonemes.

The theory of the hydrodynamic effect of mastigonemes is based on the insight that if a thin cylinder (such as a mastigoneme or any region along a flagellum) is moved through a fluid, it produces about twice as much drag on the surrounding fluid (and requires about twice as much force) at a given speed when its axis is perpendicular to the direction of movement than when its axis is parallel to the direction of movement (Hancock, 1953). In a flagellum propagating periodic planar waves, any small segment along the flagellum moves through a cyclic path, with some movement toward the base and some toward the tip during each cycle. It turns out that for typical flagellar waveforms with the waves passing from base to tip, on balance, any segment is predominantly perpendicular to its movement when going toward the tip and predominantly parallel to its movement when going toward the base. The flagellum thus produces a net thrust on the surrounding fluid toward the tip, and the cell is moved in the opposite direction (Gray and Hancock, 1955).

Taylor (1952) pointed out that a rough cylinder (i.e., with projections perpendicular to its surface) can produce thrust in the opposite direction and calculated the effect for some conditions. Jahn et al. (1964b) applied this idea qualitatively to mastigonemes: If a mastigoneme is perpendicular to a flagellum in the plane of beating, then the mastigoneme axis will be perpendicular to the direction of movement when the flagellar axis is parallel to the direction of movement, and vice versa. Hence the directions in the argument in the preceding paragraph will be reversed for the mastigoneme, and it will produce a net thrust in the opposite direction to that produced by the flagellum. A flagellum with mastigonemes of sufficient length and number can propel the cell in the direction opposite to that produced by a smooth flagellum. Holwill and Sleigh (1967) developed this argument quantitatively for mastigonemes; they compared measured and calculated swimming speeds for O. *malhamensis* and an unidentified chrysomonad and found them to be in good agreement. Brennen (1976) calculated the effects under a variety of assumptions and applied them to the observed swimming of *Ochromonas*. He obtained good agreement between the calculated speed of 60 μm s^{-1} and the observed speeds of 50–60 μm s^{-1}. His calculations indicate that significant degradation of speed would occur if the mastigonemes were not within about 10° of perpendicular, which imposes strong demands on both their stiffness and the rigidity of their connection to the flagellum.

The preceding arguments indicate not only that mastigonemes must be sufficiently long and numerous, but also that they must project into the beat plane and be held reasonably perpendicular to the flagellum during beating. In the electron microscope the mastigonemes of O. *danica* are seen to lie in two opposing rows in the plane of the centra pair (Bouck, 1971). However, the plane of the central pair is not a reliable indicator of the plane of beating (e.g., Gibbons, 1961; Satir, 1963; Tamm and Horridge, 1970; Omoto and Kung, 1980) and has been reported to vary with respect to the doublets in some algae (e.g., Herth, 1982; Melkonian, 1982; Melkonian and Preisig, 1982; Melkonian et al., 1982). It is also difficult to demonstrate directly that they are perpendicular to the flagellum during beating. However, their thick tubular construction (Bouck, 1971) and their attachment to axonemal doublets at their base (Markey and Bouck, 1977) suggest that they are rigid and that they are designed to withstand viscous forces. Because of their small size their orientation on swimming cells cannot be determined precisely in the light microscope. However, Holwill and Peters (1974) found that, in contrast to smooth flagella, fluid flow was not measurable with 1 μm of the long flagellum of O. *danica* in the plane of beating, suggesting that their mastigonemes—which are about 1 μm long (Bouck, 1971)—are in that plane and reasonably perpendicular to the flagellum.

Synurophyceae

The flagellar structure of the Synurophyceae is similar to that of the Chryso-phyceae, with which they were classified until recently (Andersen, 1987). They produce flagellate gametes that closely resemble the vegetative cells (Wawrik, 1960; Bradley, 1966; Sandgren and Flanagin, 1986). The cells have a mastigoneme-bearing flagellum and a smooth one; the smooth one is shorter in some species. However, they are inserted into the cell approximately parallel to one another, and one or both flagella have a covering of small scales (Schnepf and Deichgräber, 1969; Hibberd, 1973; Zimmermann, 1977).

Each cell of the colonial chrysophyte *Synura* bears two flagella, emerging from the same location. Jarosch (1970) has studied the motility of the vegetative cells of *S. bioretti*. One flagellum was about 40 μm long; the other flagellum was often somewhat shorter. The longer flagellum had planar "meander- or S-like" waves that moved rapidly from the base toward the tip (cf. Brokaw and Wright, 1963; Brokaw, 1965; Silvester and Holwill, 1972; see the following discussion of Dinophyceae). The shorter flagellum curved toward the cell proximally and was somewhat straightened distally. It had small helical waves that moved slowly from the base toward the tip; their amplitudes appeared quite small in the distal half of the flagellum. The helical nature of the waves of the shorter flagellum was not distinguishable in the published photographs. Quantitative data on the bend amplitudes and beat frequencies were not given. Some planar flagella had a fine fibril projecting from their tip. This fibril rotated at the same frequency as that of the flagellar beating (Fuchs and Jarosch, 1974; Jarosch and Fuchs, 1975). When cells of *Synura* were flattened by pressing on the cover slip, the waves on the planar flagellum changed in several ways (Jarosch, 1972; Fuchs and Jarosch, 1974). The frequency diminished appreciably, and the waves on the distal portion became smaller (in both amplitude and wavelength) and helical. Strongly compressed cells lost their flagella. Very slowly beating flagella stopped immediately on separating from the cell body, whereas rapidly beating flagella removed with quick pressure continued to beat for up to a few seconds.

The two flagella of the gametes of the Synurophyceae resemble those of the vegetative cells (Wawrik, 1960; Bradley, 1966; Sandgren and Flanagin, 1986). Like the vegetative cells, they swim with their flagella in front. The smooth flagellum exhibits little or no motion. Their motility has not been analyzed.

Prymnesiophyceae (Haptophyceae)

The Prymnesiophytes have both a nonflagellate and a flagellate stage. In the flagellate stage they have two flagella and often a haptonema. In the order

Pavlovales the flagella are unequal in length; the longer flagellum bears fine hairs or scales, and the shorter may bear fine hairs. These hairs are not rigid like mastigonemes; their hydrodynamic effects are not understood. In the other orders the two flagella are smooth and approximately equal in length; in some cells they are described as "subequal," with one being slightly longer. Both flagella emerge from one end of the scaly, elongate cell body, often slightly off center; the haptonema can emerge between them.

In *Pavlova* (Pavlovales) the longer flagellum is densely covered with hairs; this covering is uniform around the organelle (van der Veer, 1969; Green and Manton, 1970; Green, 1973, 1976; Green and Hibberd, 1977). The hairs are very fine and are not homologous to the mastigonemes of the Chrysophyceae. In addition, its surface is covered with small, elongated bodies (Green and Manton, 1970; Gayral and Fresnel, 1979; Green, 1980). The hydrodynamic effects of the hairs and knobs are not known. However, some workers have suggested that hairs may increase the effective diameter of a flagellum, and thereby increase the efficiency of propulsion of the cell body. The longer flagellum appears to be about 10–20 μm in the photographs of *P. gyrans* of Green and Manton (1970) and about 15–20 μm in the photograph of *P. mesolychnon* of Green (1976). The shorter flagellum may be bare or covered with fine hairs similar to those of the longer one. It appears to be about 3–5 μm in the photographs of *P. gyrans* of Green and Manton (1970); it is reduced to a vestigial structure in *P. mesolychnon* (Green, 1976).

The longer flagellum of *Pavlova* is directed forward during swimming (van der Veer, 1969; Green and Manton, 1970; Green, 1973, 1976). When at rest, it exhibits two or three very deep bends rather than being straight. In the flash photograph of a *P. gyrans* at rest in Green and Manton (1970), Fig. 2, Pl. I, the bends of the anterior flagellum appear circular. The middle and outer bends both subtend more than 4 rads., whereas of course sine-wave bends cannot subtend more than π rads. The resting waveform thus appears to be well approximated by circles and straight lines or meanders (cf. Brokaw and Wright, 1963; see the following discussion of Dinophyceae). Green and Manton (1970) describe the swimming as rapid, with the anterior flagellum having an S waveform similar to the resting one. The waveform envelope shown in their Fig. 1, Pl. I, is consistent with that shape. *P. mesolychnon* has a similar waveform (Green, 1976). According to Green and Manton (1970), the short flagellum of *P. gyrans* "is directed outwards and somewhat backwards; it beats with a stiff jerky action." Detailed descriptions have not been made of either the longer or the shorter flagellum of *Pavlova,* and the contribution of the shorter flagellum to the cell's movement is not known.

A number of authors have given qualitative descriptions of the swimming behavior of the non-Pavlovean prymnesiophytes. They commonly swim with both flagella in front, turning about their long axis; they can exhibit phototaxis (e.g., Gayral and Fresnel, 1983; Green and Course, 1983; Fresnel, 1986). *Prymnesium* exhibits at least two forms of motility (Billard, 1983). It swims in a straight path, turning about its long axis, with its flagella forward; this swimming is not very rapid. However, it pauses frequently, attaching by its haptonema while the flagella point backward and continue to beat.

Parke et al. (1955, 1956, 1958, 1959) and Manton and Parke (1962) have given detailed qualitative descriptions of the motility of nine species of *Chrysochromulina*. All the species exhibited phototaxis, and cells usually rotated during swimming. They had a haptonema that could be extended or coiled. The flagella and haptomena emerged from the cell ventrally, fairly close to one another, the distance from the region of emergence to the cell margin depending on the species. The length of the flagella varied from 10–15 μm to 30–40 μm, depending on the species. In most of the species the flagella had equal lengths, although they were sometimes slightly subequal in *C. alifera, C. chiton,* and *C. strobilus.* The flagella were usually homodynamic in most species. However, they were usually heterodynamic in *C. ephippium* and *C. alifera;* in *C. strobilus* they were homodynamic during rapid movement and heterodynamic during slow movement. In homodynamic movement, both flagella undulated; in heterodynamic movement, one undulated and the other was stationary and appeared stiff. During more rapid swimming in most species, the haptonema and flagella were behind the cell, as shown in Figure 4.2a. This was the most commonly observed pattern. The haptonema could exhibit various degrees of coiling; cells swam most rapidly when the haptonema was most tightly coiled. During slower swimming in most species—or when the cell was attached to a substrate by its haptonema—the flagella curved back along the cell and extended behind it, as shown in Figure 4.2b. In cells swimming as shown in Figure 4.2b, the haptonema was generally fully extended and appeared stiff; in attached cells it could

Figure 4.2. Some flagellar configurations seen in some species of *Chrysochromulina*. Cells moving toward top of figure. See text for details. Dorsal view. (Adapted from Parke et al., 1955, 1956, 1958, 1959 and Manton and Parke, 1962.)

be coiled. In *C. brevifilum* the flagella could sweep rapidly from the position shown in Figure 4.2b to the one shown in Figure 4.2a (i.e., they could straighten rapidly). In *C. ericina,* they could flick rapidly between these positions in either direction, causing rapid changes in swimming direction. In *C. ericina* and *C. ephippium,* cells could stop quickly by rapidly swinging the flagella out from the body. Some exceptions to the patterns shown in Figures 4.2a and 4.2b were noted. During rapid swimming, *C. ephippium* and *C. alifera* had their haptonema and straight flagellum pointing back, with their motile flagellum undulating, as shown in Figure 4.2c. During slow gliding, *C. ephippium, C. alifera,* and *C. strobilus* had their haptonema pointing forward and their flagella pointing backward, as shown in Figure 4.2d. During rapid swimming, *C. strobilus* swam as shown in Figure 4.2e. *C. polylepsis* could rapidly change the positions of its flagella to change direction and could stop by curling the ends of its flagella. Other configurations were also observed under various conditions of motility.

Moestrup and Thomsen (1986) found that *Chrysochromulina apheles* exhibited several types of swimming behavior, some of which were slow enough to analyze visually. There were two basic types of slow swimming, both occurring with the haptonema completely extended. In the first type the cells swam with the haptonema pointing forward and the flagella pointing slightly backward but almost perpendicular to the direction of swimming. The flagella beat slowly, and the cells did not rotate. In the second type both flagella made an acute angle with the haptonema, so that the haptonema and both flagella pointed backward. The flagella beat slowly, and the cell body rotated slowly about its long axis; the cells swam at 5–20 μm s^{-1}. The second type could change abruptly to a more rapid type of swimming, as the haptonema coiled tightly and the flagellar beat frequency increased. The cell continued to swim forward and rotate; however, the swim path became less straight and appeared to be a wide spiral. Swimming was generally too rapid to be analyzed from videotapes, but a speed of 160 μm s^{-1} was measured from one cell.

The various flagellar waveforms and beat frequencies have not been described in any detail in the Prymnesiophyceae. The descriptions of swimming of non-pavlovalean prymnesiophytes are all consistent with passage of waves from base to tip. The sketches of live cells suggest that their flagella are a few wavelengths long, with waves of fairly small amplitude. They have a striking ability to change their waveforms, so that the flagella either point forward or curve and point backward. These waveform changes resemble those seen in some of the Chlorophyceae, described later. However, these prymnesiophytes continue to undulate when curving backward, rather than switching to a breast stroke as the chlorophytes do. Their apparent ability to change beat frequency rapidly is also of interest.

Xanthophyceae (Tribophyceae)

Most flagellate cells in the yellow-green algae have a longer flagellum, which bears two opposing rows of mastigonemes, and a shorter smooth flagellum (Leedale et al., 1970). The longer one is directed anteriorly and the shorter one is directed posteriorly.

In the Heterogloeales, the unicellular vegetative stage can be flagellate, with the typical flagellar form (Pascher, 1925; Ettl, 1978). In the other orders the vegetative cells are nonflagellated, but often produce flagellate zoospores or gametes. The zoospores display the familiar pattern of two flagella originating subapically, with the longer, mastigoneme-bearing flagellum extending anteriorly and the shorter, smooth one curving back and pointing posteriorly (Mosto, 1978; Darling et al., 1987).

Vaucheria produces flagellate zoospores and spermatozoids. Mature spermatozoids have been described by Moestrup (1970a). They had an elongate cell body, with two laterally inserted flagella. The front flagellum bore two rows of mastigonemes. However, it was shorter than the smooth posterior one; their average lengths were 9.0 μm and 12.2 μm, respectively. He found that "when first liberated the spermatozoids swim around vigorously with one flagellum projecting forward and moving in a spiral, the other pointing backward along the body showing little or no movement." The zoospores are multiflagellate, being covered with pairs of flagella (Greenwood et al., 1957). The average lengths of the longer and shorter flagella, as measured by Greenwood (1959) under a uv microscope, were 11.1 μm and 9.8 μm, respectively. The free-swimming stage lasted about 10–15 min. Greenwood et al. (1957) noted that they swam intermittently. When the cells were stationary, their flagella still beat, albeit more slowly, and gently touching the spores could induce the spores to swim again. Stops became more frequent and prolonged; finally, rhythmic beating ceased, and the flagella straightened and projected approximately normal to the cell surface before being resorbed.

Chemotaxis has been reported in *Vaucheria* (e.g., Maier and Müller, 1986).

Neither the flagellar waveforms nor the coordination of the zoospore flagella have been described in the Xanthophyceae.

Eustigmatophyceae

The Eustigmatophyceae have been separated from the Xanthophyceae on the basis of several structural features (Hibberd and Leedale, 1970), although the justification for that separation has been disputed (Dogadina, 1986). In most (but not all) of the eustigmatophytes, the zoospores have only the mastigoneme-bearing flagellum. The other flagellum is missing, although the second basal body is present. The flagellum is directed anteriorly during swimming, as it is in the biflagellate eustigmatophytes (Lee and Bold, 1973). The details of flagellar motility in the eustigmatophytes have not been analyzed.

Bacillariophyceae

The vegetative cells of diatoms are not flagellate, although it is possible to get the impression that they are ciliated (Hogg, 1855). However, the gametes of centric diatoms are flagellate (von Stosch, 1951). They have a single flagellum (and only one basal body), although occasional biflagellate ones have been seen under some growth conditions (Schultz and Trainor, 1968). The flagellum bears two opposing rows of mastigonemes (Manton and von Stosch, 1966; Heath and Darley, 1972).

Manton and von Stosch (1966) report that the gametes swim with their flagellum in front. Although they do not give a verbal description of the waveform, the flash photographs of live cells in their Fig. 4 show the flagella to have a meanderlike waveform with bends of large amplitude. The photographs were taken with an objective lens of n.a. = 1.0, and the flagella appear to be in focus over their entire length, suggesting that the waveform is planar.

These motile flagella are unusual in that they lack a central pair of microtubules; they have a 9 + 0 ultrastructure (Manton and von Stosch, 1966; Heath and Darley, 1972). The effect of the lack of central microtubules on their motility is not known. Motile flagella with a 9 + 0 structure are fairly rare. However, several motile 9 + 0 animal gametes are known (Afzelius, 1962; Desportes, 1966; Costello et al., 1969; Baccetti et al., 1979; Gibbons et al., 1985), and the motile gametes of gregarine protists can lack central microtubules and some outer doublets (Schrével and Besse, 1975; Prensier et al., 1980; Goldstein and Schrével, 1982). In addition, the 9 + 1 flagella of the gametes of Golenkinia (Moestrup, 1972) and some 9 + 0 mutants of Chlamydomonas (Goldstein, 1982; Luck et al., 1982; Brokaw and Luck, 1985), both discussed later, are motile. Why so few flagella have dispensed with the central pair of microtubules, why they occur in such diverse organisms, and why all the known examples are gametes remain unanswered questions.

Dinophyceae

As a group, the dinoflagellates are perhaps the fastest algal swimmers (Raven and Richardson, 1984), attaining speeds of 200–500 μm s^{-1} (Peters, 1929; Hasle, 1950, 1964; Hand et al., 1965; Eppley et al., 1968; Throndsen, 1973;

Hand and Schmidt, 1975), although some other algae fall within this range (Throndsen, 1973).

The Dinophyceae are biflagellate. The flagella can insert apically ("desmokonts") or ventrally ("dinokonts").

In the Prorocentrales, which are desmokonts, both flagella emerge from the anterior end. One flagellum extends anteriorly, and the other is coiled and lies roughly perpendicular to the first one, spiraling somewhat posteriorly along the cell body (e.g., Dodge and Bibby, 1973; Soyer et al., 1982). They do not contain thecal grooves for the flagella and may be primitive dinoflagellates (e.g., Taylor, 1980; Soyer et al., 1982). Soyer et al. (1982) found that the coiled flagellum of *Prorocentrum micans* lay within a membrane; the membrane was attached to the cell body except near the tip. They observed the motility of the flagella in cells that had been slowed with 5 μl/ml of Xylocaine. They found that the coiled flagellum of the slowed cells "rotates as if on a ball-and-socket joint, with a syncopated beat, and the movement is transmitted to the differentiated part in waves, actuating the undulating membrane and then the free tip, which beats like a whip." The syncopation they saw may have been an artifact of the Xylocaine. They found that the longitudinal flagellum of the slowed cells, which is smooth, "beats with an anterior–posterior whipping action." In old stationary-phase cultures the cells commonly had two anterior flagella in addition to the coiled one; the anterior ones beat synchronously. *Prorocentrum* is unusual among algae in being pulled by means of a tip-to-base wave on a smooth flagellum, instead of by a base-to-tip wave on a mastigoneme-bearing flagellum. Indeed, flagella with proximally directed waves are rare in general. However, proximally directed waves are produced by the flagella of some nonalgal protists (e.g., Afzelius, 1962; Walker and Walker, 1962; Holwill, 1965). The motility of the flagella of the Prorocentrales has not been studied in detail.

Most dinokont dinoflagellates have a posterior flagellum, whose proximal portion runs along a longitudinally aligned groove called the "sulcus" and extends past the cell; they also have a transverse flagellum, which is located in a transverse groove called the "girdle" or "cingulum" (see Fig. 4.3a). The flagella emerge from the cell body near one another, on the ventral side.

Figure 4.3. (a) Dinokont dinoflagellate, showing flagella. (b) Transverse flagellum and strand. p = posterior flagellum; s = strand; t = transverse flagellum. (b after Berdach, 1977).

The transverse flagellum curves around the cell body in a counterclockwise direction (going from base to tip) as seen by a viewer in front of the cell. The outer edge of the transverse flagellum typically projects out of the cingulum. The transverse flagellum is about twice as long as the posterior one in some species (Leadbeater and Dodge, 1967a; Höhfeld et al., 1988).

Metzner (1929) reviewed the earlier literature and reported original observations on several dinokonts. Using India ink to visualize water currents, he reported that the longitudinal flagellum produced forward thrust and that the transverse flagellum produced both forward and rotational thrust. Jahn et al. (1963b) and Berdach (1977) also noted backward movement of particles suspended in the medium near the transverse flagellum, indicating that it produced posterior thrust. Jahn et al. (1963b) found that *Ceratium* swam forward with or without the posterior flagellum beating. From observations of cells with a long, a short, or no posterior flagellum, Hand and Schmidt (1975) concluded that the posterior and transverse flagella of *Gyrodinium dorsum* each provide the thrust for about half of the forward speed. The cells rotated at an average rate of about 0.7 s per turn when swimming; this rate was not appreciably altered by loss of the posterior flagellum. *Noctiluca*, which has a greatly reduced transverse flagellum, does not rotate while swimming (Zingmark, 1970). The relative forward thrust provided by each of the two flagella may vary with species. Gaines and Taylor (1985) surveyed over 50 species. They reported that in a few species (e.g., *Ceratium*) the posterior flagellum appeared to provide most of the forward thrust and turning was minimal, whereas in most species the posterior flagellum appeared to act primarily as a rudder and to provide relatively less thrust, but they did not report rates of rotation or advancement.

The portion of the posterior flagellum projecting past the sulcus can be over 200 μm long, making it convenient for waveform measurements. It beats with planar undulations. Metzner (1929) found the beat plane in *Peridinium* to be perpendicular to the ventral and dorsal surfaces of the cell body; Jahn et al. (1963b) found them to be parallel in *Ceratium*, as shown in Figure 4.3a. In *G. dorsum* the posterior flagellum beats at about 60–65 Hz (Hand and Schmidt, 1975). At 20°C, the posterior flagellum of *C. tripos* beats at 30 Hz and has a wavelength of about 74 μm and an amplitude of about 14 μm (Maruyama, 1981). Brokaw and Wright (1963) noted that in *Ceratium* the peaks of the waveform of the posterior flagellum are more rounded than sine waves. The waveform could be better approximated by a series of circular arcs connected by straight-line segments. This sublety does not greatly affect the efficiency of the flagellum in propelling the cell; however, it does have interesting implications for mechanisms and models of beating (e.g., Brokaw, 1965, 1968, 1983; Hines and Blum, 1978, 1979). Once this form has been pointed out, it is obvious in published photographs of many algal flagella. It is especially noticeable in flagella such as the anterior one of *Pavlova gyrans,* mentioned earlier (Green and Manton, 1970), which form large fractions of a complete circle.

Bends may not be precisely circular in some flagella. A series of circular and straight regions closely resembles another type of S-shaped curve, called a "meander" (Silvester and Holwill, 1972). Meanders are encountered in various situations in mechanics; for instance, a uniform beam whose ends are separated by less than its length forms a meander. Such a curve is clearly of interest in considering mechanisms of flagellar bending. However, distinguishing between these two types of S-shaped curves in photographs of flagella is difficult and can require comparisons of waveform harmonics with Fourier analysis (Silvester and Holwill, 1972). The limited resolution inherent in light micrographs can require this analysis to be done as an average over many images (Johnston et al., 1980; Marchese-Ragona et al., 1983). Another feature can be seen in some published photographs of bends of the posterior dinoflagellate flagellum: the radius is not constant within a bend, but increases somewhat from its proximal to its distal end (e.g., Jahn et al., 1963b; Maruyama, 1981). This increase in radius has also been noted in flagella of nonalgal cells (e.g., Goldstein, 1969).

The shape of the transverse flagellum has proven more difficult to analyze than that of the posterior one. This flagellum and an adjacent "strand" are contained within a membrane (e.g., Berdach, 1977). The strand is generally striated, although Rees and Leedale (1980) found the strand of *Peridinium cinctum* to appear amorphous. On the basis of light-microscope observations, Kofoid and Swezy (1921) described the shape of the flagellum and associated membrane as "ripples or folds." Hall (1925), Metzner (1929), Peters (1929), and Jahn et al. (1963b), also using light microscopy, considered the lateral flagellum of *Ceratium* to be helical. In early studies using the scanning electron microscope, it was sometimes considered helical (Leadbeater and Dodge, 1967a,b), and it sometimes appeared to be ruffled rather than helical (Taylor, 1975; Herman and Sweeney, 1977; Soyer et al., 1982). However, in a comparison of the effects of several fixation and dehydration procedures on the lateral flagellum of *P. cinctum,* Berdach (1977) showed that preparation for scanning electron microscopy can cause variable amounts of distortion of the lateral flagellum and concluded that its natural form is approximately helical. However, it is not a simple circular helix. The side of the helix farthest from the cell body is angled more parallel to the anterior–posterior axis of the cell than the rest of the helix, so that it resembles a coil in which each turn has been stretched within the cingulum and compressed where it is more exposed to the medium; that is, the pitch angle is greater within the cingulum than at the peripheral edge of the helix (see Fig. 4.3b). This modified helical shape now appears to be the accepted form of the transverse flagellum (e.g., Rees and Leedale, 1980; Gaines and

Taylor, 1985; Höhfeld et al., 1988). The helix is left-handed. Pitches typically fall in the range of 3–6 μm (Gaines and Taylor, 1985). The pitch in *Scrippsiella trochoidea* is about 4 μm (Höhfeld et al., 1988).

There has also been some confusion about the relationship between the transverse flagellum and the adjacent strand, which runs parallel to the axis of the flagellum. Leadbeater and Dodge (1967a,b) described it as lying within the helical flagellum. However, as Berdach (1977) has shown, it runs between the inner edge of the flagellum and the wall of the cingulum. The transverse flagellum and strand lie within a membranous covering, which appears to be tightly wrapped around them. Fine hairs project from the outer edge of the flagellum (Leadbeater and Dodge, 1967a,b; Rees and Leedale, 1980). In the electron microscope the structure of these hairs appears to be more like that of the flexible type than like mastigonemes.

Waves pass along the helical flagellum. Videorecorder observations by Gaines and Taylor (1985) indicate that they go in a counterclockwise direction as seen by a viewer in front of the cell; that is, they pass from the cell's right to its left on the ventral side. (They are *leiotropic,* as opposed to *dexiotropic.*) They also found that each point on the transverse flagellum appears to rotate clockwise when viewed from the base toward the tip. That is, the transverse flagellum appeared to be a left-handed helix passing waves from its base toward its tip. They do not give rotation frequencies; however, the flagellum traced in their Figure 10 appears to be rotating roughly five times per second.

Reports of the direction of rotation of the cell body during swimming vary. Kofoid (1906) wrote that "in the Dinoflagellates generally the rotation is predominantly from left over to right," which Jahn et al. (1963b) have interpreted as meaning leiotropic and Gaines and Taylor (1985) have interpreted as meaning dexiotropic. Metzner (1929) and Gaines and Taylor (1985) reported that it was predominantly leiotropic. Peters (1929) and Jahn et al. (1963b) reported that the direction of rotation could switch between dexiotropic and leiotropic. Sleigh (1981) has suggested that the transverse flagellum may be able to reverse the direction of wave propagation. However, Gaines and Taylor (1985) did not find cells capable of switching to dexiotropic rotation. They believe that previous reports of reversals are erroneous and due to the transparent cells swimming through the focal plane. Because waves on the ventral and dorsal surfaces pass in opposite directions across the cell, a sudden switch of focus between these two surfaces can give the illusion of a reversal of beating, especially at low magnifications.

The study of Gaines and Taylor (1985) is especially interesting because it attempted to correlate the direction of rotation of the cell with the direction of beating of the transverse flagellum. They found that both were leiotropic; that is, the cell rotated in the same direction as the flagellar waves were passing. This is a rather surprising result. The most straightforward expectation would be that on, say, the ventral side, flagellar waves passing to the cell's left would push the surrounding water to the left, and the cell (against which the flagellum is pushing) would be rotated in the opposite direction. In this regard it may be of interest that Peters (1929), in his Figure 10, shows the transverse flagellum of *P. claudicans* extending to the cell's left and the swim path as a right-handed helix. The cell maintained its ventral side toward the helix axis, which implies that—if waves were passing distally on the transverse flagellum—it also rotated in the same direction as the waves were passing.

Gaines and Taylor (1985) suggest that the reversal of torque from the expected direction is due to the hairs on the outer edge of the flagellum, which act as mastigonemes (see the preceding discussion of the Chrysophyceae). Although mastigonemes can reverse the propulsive force of helical waves (Holwill, 1966a), there are difficulties with this suggestion. As they point out, the electron-microscope structure of the hairs in dinoflagellates resembles the fine flexible hairs of other algae rather than that of rigid tubular mastigonemes. In addition, it is not clear that the putative mastigonemes would undergo enough rotation in the direction of passage of the flagellar waves (i.e., parallel to the helix axis) to cause a significant hydrodynamic effect. The flagellum itself does not actually rotate, of course; the progression of waves along it reflects the sliding of its doublet microtubules (e.g., Satir, 1965; Summers and Gibbons, 1971; Brokaw, 1989). The doublets may undergo very little torson in the direction of the helix axis while they are sliding. At the outer edge of the helix, where the flagellum is most exposed to the medium and mastigonemes would be most extended and separated (and hence probably most effective), the flagellum is fairly perpendicular to its helix axis. So the movement of the mastigonemes due to sliding of the doublets might be largely perpendicular to the helix axis where they would probably be most effective, rather than parallel to it. There are three other features of the transverse flagellum that no doubt modify its hydrodynamics and might play a role in reversal of rotation: (1) The flagellum is not a uniform helix; its helix angle is larger next to the cell than at the outer edge of the cingulum. (2) A large part of each turn is located within the cingulum, which presumably affects water flow—although Gaines and Taylor (1985) noted that the swimming of *Prorocentrum,* which lacks a cingulum, is not noticeably different from that of dinokonts. (3) The flagellum is contained within a membrane whose inner edge is shaped by the strand, so that the axoneme is the outer edge of a complex surface whose traveling waves develop the torque to rotate the cell. It would be useful to have a description of the movement of this surface during the passage of flagellar waves.

The posterior flagellum of some dinoflagellates can contract rapidly (Schütt, 1895; Metzner, 1929; Peters, 1929; Afzelius, 1969; Maruyama, 1981). Maruyama (1981) has analyzed this retraction in *Ceratium tripos.* It occurred in response to mechanical stimulation of the body. It was completed within 28 ms. In viscous solutions the contraction slowed enough to be analyzed. It started at the tip and proceeded to the base. The proximal end continued to beat at first, but then stopped and straightened before the wave of retraction reached it. The flagellum folded into a long zig-zag shape, with a fold every 4–5 μm. This folded flagellum then coiled into a right-handed helix. It relaxed much more slowly, usually reextending within 1–5 s at room temperature. Usually the proximal end extended first, then the distal end, with the middle relaxing last. The proximal end began to beat before the rest of the flagellum had extended. It contains a striated fiber (the "R-fiber") that runs parallel to the axoneme. This fiber contracts upon increase in free Ca^{2+} concentration (Maruyama, 1985a,b).

The transverse flagellum of *Peridinium inconspicuum,* along with its striated strand, can be isolated from the cell (Höhfeld et al., 1988). When Ca^{2+} is added to the isolated preparation under the appropriate conditions, the striated strand contracts in a manner similar to the contraction seen in the R-fiber by Maruyama (1985a,b). This suggests that the primary function of the striated strand is contractile. It might aid in holding the transverse flagellum in the cingulum by contracting and causing the flagellum to curve toward the cell body (Taylor, 1975; Rees and Leedale, 1980; Höhfeld et al., 1988) and might serve to control the pitch or the total length of the flagellum. Gaines and Taylor (1985) report that "occasionally the fibrillar strand contracted, pulling the flagellum close to the cell surface and stopping the beat." It is not yet clear whether contraction of the striated strand directly stops the flagellar beating or whether they are both responses to, say, increased Ca^{2+}.

Each flagellum can be quiescent under some conditions. The posterior and trasverse flagella of *Ceratium* can each beat while the other one is beating or quiescent (Jahn et al., 1963b). In *Gyrodinium dorsum* both flagella stop within 0.5 s of a sudden increase in light intensity (Hand and Schmidt, 1975), and when the posterior flagellum of *C. tripos* retracts, the transverse flagellum usually stops for a second (Maruyama, 1981).

Dinoflagellates can change or reverse swim direction (Peters, 1929; Hand et al., 1965; Hand and Schmidt, 1975; Herman & Sweeney, 1977; Maruyama, 1981). It involves a change in the orientation of the posterior flagellum. The flagellum stops beating, points in a more anterior direction by bending (mainly where it exits the sulcus), and resumes beating with its usual waveform over most of its length. It may remain stopped for several seconds before resuming beating. Maruyama (1981) reported that, in addition to exhibiting the usual reversal behavior, *Ceratium tripos* can "dash" backward when the posterior flagellum is kept retracted; no mechanism was suggested for this movement.

Dinoflagellates, like most flagellate algal cells, usually rotate while swimming and exhibit phototaxis. In response to a sudden increase in light intensity from one direction, *Gymnodinium splendens* and *Gyrodinium dorsum* stop for up to several seconds and then swim toward the light source (Forward and Davenport, 1970; Forward, 1974). Both species exhibit a circadian rhythm of sensitivity to light. Hand and Schmidt (1975) have examined the behavior of the posterior and transverse flagella of *G. dorsum* during such a phototactic response. Cells stopped within 0.5 s of the stimulus. The posterior flagellum stopped, and the transverse one also appeared to stop. The posterior flagellum bent toward the light source. It was gently curved distally and bent most sharply where it exited the sulcus. It could remain quiescent for several seconds. It then resumed planar beating. The angle between the cell axis and the flagellum changed as the cell reoriented, until it extended posteriorly again as the cell swam toward the light.

Dinoflagellates can produce motile gametes. *Gymnodinium pseudopalustre* and *Woloszynskia apiculata* produce isogametes that resemble the vegetative cells and have transverse and posterior flagella (von Stosch, 1973). The planozygote that results from fusion of the gametes has four motile flagella. The two posterior flagella appear in his Figure 10 to be beating in synchrony, with a typical waveform. *Noctiluca miliaris* produces isogametes with a single, posterior, flagellum that beats "with sinuous waves" (Zingmark, 1970). After fusion, both flagella continue to beat, but then slow down, stop, and shorten. The flagellar motility of dinoflagellate gametes has not been studied in detail.

Xiaoping et al. (1989) have observed a remarkable behavior of the transverse flagellum in gametes of *Scrippsiella*. During mating, "the transverse flagellum of one gamete migrates out of the girdle and grasps the longitudinal flagellum of the other gamete and then returns to the girdle. . . ." The gametes are thereby united and fusion ensues. The strand of the transverse flagellum is an obvious candidate as a source of this movement. After fusion is complete, a motile zygote is formed that has two longitudinal flagella and one transverse one.

Phaeophyceae

Vegetative cells of the brown algae are not flagellate. However, all brown algae produce motile gametes and/or zoospores. In the Dictyotales they have two basal bodies but only an anteriorly directed, mastigoneme-bearing flagellum (e.g., Manton, 1959a). In all the other orders they have an anteriorly directed mastigoneme-bearing flagellum and a posteriorly directed smooth one. In the Fucales the posteriorly directed one is longer (e.g., Manton and Clarke, 1956). In the remaining orders the anteriorly directed one is generally longer (e.g., Manton and Clarke, 1951; Henry and Cole, 1982), although in the Laminariales the posterior flagellum can have a very long, thin terminal extension that makes its total length greater than that of the anterior flagellum (e.g., Henry, 1987).

The appearance of the mastigonemes is somewhat different from those in other classes. Although they have usually been described as lying in two opposing rows (e.g., Manton et al., 1953; Henry and Cole, 1982), they probably lie in a group on one side of the flagellum (e.g., Bouck, 1969; Loiseaux, 1973; Toth, 1976). Although they have a tubulelike structure, they are attached to the membrane and show no sign of being anchored to the axonemal doublets (Bouck, 1969). Their role in motility is not known. The flagella of some species have one or more spines protruding from their side (see Moestrup, 1982, for review); their function and effect on motility are not known.

Swimming involves beating of the anterior flagellum, but not the posterior one (Toth, 1976; Geller and Müller, 1981; Henry and Cole, 1982; Müller et al., 1985a). During swimming, the anterior flagellum of the zoospore of *Chorda tomentosa* "generates a wave-like motion and the posterior flagellum remains rigid" (Toth, 1976). The anterior flagellum contains a knoblike tip, which is actually a highly coiled region about 20 μm long, which attaches to the substrate when the cell settles and the flagella are withdrawn into the cell body. Henry and Cole (1982), observing 17 species of the Laminariales, found that the anterior flagellum, but not the posterior one, "beats sinusoidally." A meanderlike wave can be seen on the anterior flagellum of a live *Perithalia caudata* spermatozoid in Figure 19 of Müller et al. (1985a); the posterior flagellum trails passively. The anterior flagellum of the male gamete of *Ectocarpus siliculosus* produces both planar meanderlike undulations and three-dimensional waves (Geller and Müller, 1981). They found that gametes swimming far from a surface "exhibit an irregular beat of their front flagellum. . . . Planar to elliptical waves alternate with rotation of the front flagellum." Next to a surface the waves were planar. The hind flagellum did not beat.

Male gametes of a number of brown algae have been shown to exhibit chemotactic attraction to female gametes (e.g., Cook et al., 1948; Cook and Elvidge, 1951; Geller and Müller, 1981; Maier, 1984; Müller et al., 1985a,b; Maier and Müller, 1986; Henry, 1987). They respond to species-specific hydrocarbon pheromones (e.g., Müller, 1978; Müller and Lüthe, 1981; Müller et al., 1985a,b; Maier and Müller, 1986). These pheromones can also induce release of the male gametes. The response to pheromones varies with species (Maier and Müller, 1986). *Laminaria digitata* approaches a source of pheromone rather directly, from distances of at least 1 mm. When heading away from a source, *Fucus spiralis* performs a U turn when it passes through a critical concentration and swims back toward the source. *Ectocarpus siliculosus* swims in loops of decreasing radius as the concentration increases, so that it spirals more or less directly toward a source.

Chemotaxis has been studied in some detail in *E. siliculosus,* using video (Müller, 1978) and stroboscopic photographs and high-speed film (Geller and Müller, 1981). The anterior flagellum was 12 μm long, with a terminal filament that was 10–13 μm long (Müller and Falk, 1973) and wound into a spiral. The posterior flagellum was 4 μm long with a terminal filament 2 μm long. The pheromone in this species is called *ectocarpen.* In the absence of ectocarpen, the gametes were described as swimming in straight lines when far from a surface and in wide loops when in contact with a surface. Viewed from above, they swam clockwise against the cover glass and counterclockwise against the slide. However, their curved swim paths against the glass suggest that their paths when not near a surface may be somewhat helical rather than straight: cells swimming in helical paths tend to swim in curves against a surface when they encounter one (cf. Gray, 1955). The clockwise swim direction of *E. siliculosus* suggests that they may swim in a left-handed helical path when not hindered by a surface.

Müller (1978) found that in the absence of ectocarpen, the average swim path radius (r) of *E. siliculosus* was 137 μm. In the presence of 10^{-3} M ectocarpen they swam in circles 50 μm in diameter and often made sharp turns. When swimming in the vicinity of an egg, they could turn too sharply and miss the egg. They swam more slowly against a glass surface than when away from a surface, but swim speed was independent of the presence of ectocarpen (Geller and Müller, 1981). However, the average speed decreased with turn radius, going from about 110 μm s^{-1} when $r \geqslant 50$ μm to about 60 μm s^{-1} when $r = 5-15$ μm at 19°C in the presence of a female gamete. Average beat frequencies of the front flagellum decreased somewhat with swim path radius, being about 44 Hz when $r \geqslant 50$ μm and about 31 Hz when $r = 5-15$ μm. Geller and Müller (1981) do not explicitly state that the waves were directed distally. The decreasing swim path radius reflected changes in both the anterior and posterior flagella, as shown in Figure 4.4. The anterior flagellum became more asymmetric. They found a linear

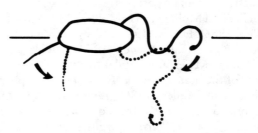

Figure 4.4. Chemotactic turning in *Ectocarpus siliculosus*. Cell swiming toward right. Solid lines = flagella when swiming in wide circles; dotted lines = flagella when swiming in small circles. Arrows indicate direction of deflection of flagella to reduce swim path radius. (After Geller and Müller, 1981.)

correlation between the deflection of the most proximal visible bend (computed as the area between that bend and the cell axis, averaged over several cycles) and the radius of the swim path. As the radius of the swim path decreased, the deflection of the posterior flagellum (measured as the angle between its tip and the cell axis) increased. The posterior flagellum could produce isolated rapid beats, and sharp turns in the swim path ($r \le 10 \ \mu m$) usually reflected a rapid beat of the hind flagellum. These beats and sharp turns were more frequent in the presence of ectocarpen. The beat waveform of the hind flagellum was not described in enough detail to tell whether waves passed along it or whether bending was confined to its base. The basal regions of both flagella are within the cell body, making their description more difficult.

Chemotaxis has also been described for bracken fern spermatozoids (Brokaw, 1957, 1958a,b, 1959) and for a number of invertebrate spermatozoa (e.g., Miller, 1966, 1970, 1975, 1977, 1979). The mechanism of turning has been investigated for some of these cells; turning is associated with transient changes in the symmetry of the flagellar waveform (Miller and Brokaw, 1970; Miller, 1976). Transient asymmetric bending, possibly associated with chemotaxis, has also been observed in fungal spores and gametes (e.g., Miles and Holwill, 1969; Pommerville, 1978; Holwill, 1982).

Cryptophyceae

The cryptophytes have two flagella, which emerge from the right side of a subapical depression, the *vestibulum,* and are aligned dorsoventrally (Kugrens et al., 1987). They are typically about as long as the cell body, with one flagellum being somewhat longer than the other. This pattern, which is typical for vegetal cells, has been seen on both haplomorph and diplomorph vegetal cells of *Proteomonas sulcata* (Hill and Wetherbee, 1986) and on gametes of *Chroomonas acuta* (Kugrens and Lee, 1988). The flagella can have mastigonemes and fine hairs, but the pattern is quite variable from species to species (e.g., Kugrens et al., 1987). In a number of species the flagella also have scales (Pennick, 1981; Santore, 1983; Lee and Kugrens, 1986).

Lowndes (1943) described specimens of a cryptophyte later identified as *Hemiselmis rufescens* (Throndsen, 1973). The anterior flagellum was about one and one-half times as long as the cell body, and the posterior one was slightly shorter. They emerged laterally; one extended anteriorly and the other one extended posteriorly. Free-swimming cells followed a spiral path whose pitch was much longer than the length of the cell. The beat frequency could not be measured and the waveform was not described.

The gametes of *Chroomonas acuta* form pairs and swim around each other, touching several times (Kugrens and Lee, 1988). The flagella are not directly involved in their fusion (Wawrik, 1971; Kugrens and Lee, 1988).

Flagellar waveforms do not appear to have been studied in the crypto-monads.

Raphidophyceae (Chloromonadophyceae)

The chloromonads typically exhibit the familiar pattern of two flagella emerging from the anterior end, with a mastigoneme-bearing one extending forward and a smooth one trailing back along the cell body. The anterior one can be either longer (e.g., Hara and Chihara, 1985) or shorter (e.g., Spencer, 1971) than the posterior one. Mignot (1967, 1976) has noted a striated appearance in the anterior flagellum of several species, which is apparently due to their mastigonemes.

Cells can spurt forward in a straight path or swim with a rotary motion (Prescott, 1968; Spencer, 1971). The anterior flagellum undulates and provides the thrust for swimming (Prescott, 1968; Spencer, 1971; Heywood, 1972; Hara & Chihara, 1985). It has been sketched as having planar mean-derlike waves in some species (Hara and Chihara, 1985), but Mignot (1976) has described the flagellum of *Chattonella subsalsa* as having a helical form and functioning in the same manner as the transverse flagellum of dinoflagel-lates (described earlier). The posterior flagellum can be fairly motionless or oscillate gently. Hara and Chihara (1985) have noted that the posterior flagellum of *Fibrocapsa japonica* does not beat, and they have suggested that it may have a steering function. Flagellar waveforms have not been analyzed in detail in the chloromonads.

Euglenophyceae

The flagella of euglenoid cells always emerge from an invagination, called a *canal*, at their anterior end. They generally have two basal bodies. In the Eutreptiales and Heteronematales, two flagella emerge from the canal. In the Eutreptiales, they are equal or nearly equal in length; one is directed anteriorly during swimming while the other is directed posteriorly or later-ally. In the Heteronematales, one flagellum is long and extends anteriorly during swimming while the other is short and is directed posteriorly. In most of the Euglenales, one flagellum is too short to emerge from the canal. In a few species of *Euglena* neither flagellum is long enough to emerge from the canal. In the other orders, one flagellum is either too short to emerge from the canal or emerges only a short distance. The only order with more than two basal bodies is the Euglenamorphales, a very small group of parasitic cells that can have three to seven emergent flagella (Brumpt and Lavier, 1924; Wenrich, 1924).

Both flagella have a *paraxial rod* running parallel to the axoneme; they have been well described in *Euglena* (e.g., Piccinni et al., 1975; Bouck and Green, 1976; Hyams, 1982; Melkonian et al., 1982). Both the axoneme and rod are contained within the flagellar membrane. The rod has been reported to have ATPase activity (Piccinni et al., 1975), and structures connect the rod to the axoneme, suggesting that the rod might play an active role in the shaping of the flagellar waveform.

Euglenoid flagellates do not have mastigonemes, but they have fine, flexible hairs (see Moestrup, 1982, for review). They are found on both emergent flagella in *Distigma* and *Eutreptiella*. In *Euglena gracilis,* distal to the canal the flagellum has two types of hairs (e.g., Bouck et al., 1978; Melkonian et al., 1982). Longer ones (~3 μm long) are arranged along one side of the flagellum and appear to be anchored to the rod (Bouck et al., 1978) or to the structures that connect the rod to axonemal doublets (Melkonian et al., 1982). Shorter ones (~1.5 μm long) are arranged helically around the flagellum and are attached to the membrane. The hairs are arranged in bundles and can be 4–5 μm long in *Eutreptiella gymnastica* (Moestrup, 1982). Leedale et al. (1965) and Leedale (in Holwill, 1966a) found the emergent flagellum of *Euglena spirogyra* to have hairs 2–3 μm long arranged in a single row distal to the canal. The hairs were fully extended in beating cells, rather than being wrapped around the flagellum as suggested by Pitelka and Schooley (1955). *Peranema* has a similar arrangement (Hilensky and Walne, 1985). The function of the hairs is not known. They may simply increase propulsive efficiency by increasing the roughness and effective diameter of the flagella, as has often been suggested, but the complexity of their arrangement may reflect more subtle effects. The longer hairs are inserted so that their tips point toward the distal end of the flagellum (see Moestrup, 1982, for review). The various observations taken together suggest that the longer hairs may form a ribbon of variable width, which splays out when a flagellum moves away from the cell body and compresses when the flagellum moves toward the cell body. This would give the flagellum a variable hydrodynamic resistance, which would be greatest when it pushes away from the cell. Because euglenoid flagella often produce net thrust away from the cell (discussed later), these hairs may be an elegant system for increasing flagellar propulsion under some conditions. The orientation of the insertion of the hairs relative to the plane of bending at any point is not known, although Bouck et al. (1978) found that the paraxial rod did not maintain a constant orientation with respect to the central pair.

Leedale (1967) has reviewed patterns of flagellar motility and given qualitative descriptions of them for species in all the orders of euglenoid flagellates. In the Eutreptiales, both flagella are highly mobile along their length; but they beat with different waveforms, one being directed anteriorly and the other usually being directed laterally or posteriorly. In the Euglenales, large waves travel along the emergent flagellum. (In those species in which both flagella are short and nonemergent, one of the flagella is generally described as motile.) In the Rhabdomonadales, the emergent flagellum is usually held straight when the cell is stationary, but it is highly mobile along its length when motile. Species in the Sphenomonadales have one or two emergent flagella. One flagellum is directed anteriorly and straight. Leedale (1967) recorded shallow waves traveling from base to tip in the anterior flagellum of *Sphenomonas laevis,* and minute waves at the tip of a straight flagellum in *Notosolenus apocamptus.* The shorter emergent flagellum was "relatively inactive, with lethargic 'cilium-like' beat" in *S. laevis;* it trails passively back in other species. The anterior flagellum is usually the longer one in this order, but it is the shorter one in *Anisomema.* Species in the Heteronematales, which includes *Peranema,* have one or two emergent flagella. One flagellum is directed anteriorly and is straight along most of its length, at least in some types of movement, with the tip region usually exhibiting waves of a coiling or flicking appearance. The other flagellum, if present, trails posteriorly. The anterior flagellum may be the longer or the shorter flagellum, depending on the species. In the Euglenamorphales, all the flagella have the same length and motility (Leedale, 1967).

The waveform of a euglenoid flagellum can be quite variable, depending on the conditions (e.g., whether the cell is gliding, swimming near to a surface, or swimming far from a surface). It is therefore important to specify the conditions under which a given waveform was recorded. Differences between different authors' descriptions of the same type of cell are probably due primarily to differences in recording conditions.

The motility of *Euglena* has been studied by a number of workers. It has been known since the beginning of the twentieth century that helical waves travel from base to tip along its emergent flagellum. The number of wavelengths on a flagellum varies with the length of the flagellum, from less than one wavelength in *E. intermedia* and *E. deses* (Jahn and Bovee, 1968) to perhaps six in *E. inflata* (Gojdics, 1953). The flagella of *E. gracilis* and *E. viridis,* the most commonly studied species, are about two wavelengths long. Early work has been reviewed by Jahn and Bovee (1964, 1968).

Leedale et al. (1965) noted that the emergent flagellum of *E. spirogyra,* which was 40–60 μm long, moved in a variety of ways for different modes of movement and direction changes. "During active swimming, the flagellum is usually projected forwards, appearing rather like a spinning lasso."

Lowndes (1936, 1941), filming cells swimming unimpeded at about 30 frames per second, with exposure times of 1/8000 s or less, found that the flagellum of *E. viridis* was 128 μm long and curved back near the base, so that there was an acute angle between the flagellum and the cell body axis. A series of waves passed from base to tip, as shown in Figure 4.5. The cell moved forward while its body axis gyrated in a conical fashion; a wobble

Figure 4.5. Waveform of emergent flagellum of *Euglena*. (After Jahn and Bovee, 1968.)

due to the generation of flagellar waves was superimposed upon the circular movement of the anterior end. From the gyrations of the cell body he concluded that the frequency of bend formation was 12.7 Hz. Wave speed was 813 μm s^{-1}, and cell speed was about 168 μm s^{-1}. Although the handedness of the waves was not given explicitly, the body rotations could be either clockwise or counterclockwise, indicating that both right- and left-handed helical bends were present in the cultures observed. Lowndes (1943, 1944) argued that the forward swimming was not due directly to the flagellar waves; rather, it resulted from the gyrations of the cell body. He presented evidence that body rotations could provide the necessary thrust, using experiments on large-scale models.

Holwill (1966a) used high-speed films to record *E. viridis* swimming unimpended in a cavity slide. The average swim speed was found to be 80 μm s^{-1}. Gyrational movements were such as to cause line segments through the cell axis to lie on a surface approximating a hyperboloid of one sheet when forward displacements were subtracted. The flagellum was directed posteriorly and lay alongside the cell body. Waves passed from base to tip, with an average frequency of 12 cycles per second. Bends began as planar at the base and became helical as they passed to the tip. The average wavelength and amplitude of the helical waveform were 35 μm and 12 μm, respectively. The handedness of the bends was not given. The flagella were about 100 μm long, or long enough for roughly two wavelengths of the preceding dimensions. Calculations of forward speeds that could be produced by these waves indicated that—contrary to the assertions of Lowndes (1943, 1944)—the flagellar waves, but not the gyrations of the cell body, were sufficient to produce the observed swimming. He noted that Lowndes' experiments were invalid because they used motions at large Reynolds numbers (and hence involved predominantly inertial forces), whereas microorganismal swimming uses motions at low Reynolds numbers (and involves predominantly viscous forces). In any simulations of flagellar movements it is important that the Reynolds numbers be in the same range as those for flagellate microorganisms ($\sim 10^{-6}$–10^{-3}).

Jahn and Bovee (1968), using tracings from high-speed films of *Euglena*, characterized the flagellar waveform as an "interrupted helix" consisting of helical bends with a "straight or almost straight" section separating the bends. They noted that the diameter of the helical bends increases as they travel distally. They showed tracings of a turning response in which the basal bend straightened so that the flagellum, still beating, was thrown anteriorly. The handedness of the helical bends is not mentioned explicitly by Jahn and Bovee (1967, 1968); both left- and right-handed bends can be found in their sketches.

Euglena responds to a number of stimuli by a change in motility. The response to at least some stimuli involves a change in the angle between the flagellum and the cell body axis. These responses have been reviewed by Jahn and Bovee (1968). When *E. viridis* touches an object the flagellum swings forward and can cause the cell to rotate about its posterior end or swim backward (Lowndes, 1936).

The effects of transmembrane voltage, intracellular bivalent ions, and light on the motility of *E. gracilis* have been investigated (Nichols and Rikmenspoel, 1977, 1978a,b, 1980; Nichols et al., 1980). They impaled cells on microelectrodes or injection micropipettes. In impaled cells held away from a surface under illumination with white (420–700 nm) light, the flagella extended anteriorly, away from the cell body, and beat with an irregular asymmetric helical waveform. That is, impaling the cell appeared to produce a prolonged reversal or avoidance response. Waveforms were monitored with high-speed film. The maximum beat frequency was 16.3 Hz in solutions containing 10 mM ATP. Displacements from a line through the cell axis were measured at 7.5 μm from the base and at the tip. Plots of displacement versus time were roughly sinusoidal; peak-to-peak values were ~20 μm at 7.5 μm and ~15 μm at the tip. Changes in conditions caused reductions in beat frequency and caused the flagellum to be directed more laterally or posteriorly. Measurements in solutions containing various concentrations of bivalent ions suggested that electrode voltage effected frequency and waveform changes by affecting the flow of Ca^{2+} and Mg^{2+} across the membrane; that is, beating appeared to respond to intracellular concentrations of bivalent ions rather than to transmembrane voltage per se. Interpretation of their data is hampered somewhat by their terminology, which is the opposite of the conventional usage for freely swimming cells. They refer to the anteriorly directed waveform as the "forward" or "normal" one, and they refer to a change to a more posteriorly directed waveform as "reversal." An increase in intracellular $[Ca^{2+}]$, a decrease in intracellular $[Mg^{2+}]$, or illumination with yellow (530–700 nm) light caused a decrease in beat frequency. In their terminology, these changes also caused an increase in the percentage of cells exhibiting waveform reversal. In conventional terms, these treatments could reduce or abolish the reversing effects of impalement. The similarity of the effects of changes in illumination and ionic conditions suggests that they are mediated by the same mechanism. Their results are certainly consistent with the hypothesis that responses to light are mediated by divalent ions in *Euglena*.

The flagellum of *E. fusca* has been seen moving in and out of the canal "with a straight back and forth motion" at times (Gojdics, 1953). Jahn and Bovee (1968) cite Chadefaud and Provasoli (1939) as reporting that flagellar retraction occurs in *E. gracilis*. However, this appears to be a mistranslation; they reported a retraction of the caudal end of the cell (*la queue*) during metabolic contractions, but did not describe movements of the emergent flagellum (*le fouet*).

Vora et al. (1967) have extracted flagella isolated from *E. gracilis* and reactivated them with solutions containing ATP. The flagella beat vigorously and moved rapidly through the medium. Suzaki and Williamson (1986) have extracted cells of *Astasia longa* with glycerol and detergent and obtained swimming similar to that observed in intact cells upon reactivation.

Peranema has two emergent flagella. The longer one is directed anteriorly during swimming, and the other one curves posteriorly and is pressed close to the cell body (e.g., Leedale, 1967). The motility of the anterior flagellum of *P. trichophorum* has been described by a number of workers. Its waveform depends on the cell's mode of movement. Holwill (1966b) has reviewed the literature on its waveforms and distinguishes three types of movement: (1) gliding along a surface; (2) movement seen in avoiding reactions; (3) free swimming in a medium.

(1) When *P. trichophorum* glides along a surface, its longer flagellum is extended anteriorly; it is straight for most of its length, with only the distal fourth or third beating (e.g., Lowndes, 1936; 1941, 1944; Chen, 1950; Jahn et al., 1963a; Holwill, 1966b; Jahn and Bovee, 1967). Descriptions of the tip waveform have varied. Verworn (e.g., 1909) thought it continued straight and vibrated with small-amplitude symmetric bends. Lowndes (1936, 1941, 1944), using films taken at about 30 fps or 60 fps, and Chen (1950), thought that the tip bent to the right to form an acute angle with the proximal portion and that small-amplitude sine waves passed distally along the tip region. However, later studies using high-speed film (Jahn et al., 1963a; Holwill, 1966b; Jahn and Bovee, 1967) have indicated that the tip beats with an asymmetric, ciliumlike beat, as shown in Figure 4.6. However, Jahn et al. (1963a) reported seeing "flagellar cycles alternating from one side to the other on either side of the vertical with an average total declination of 50°" in a marine species. (2) When *P. trichophorum* encounters an object, it backs

Figure 4.6. Waveform at tip of anterior flagellum of *Peranema trichophorum*, moving along a surface. Numbers indicate order of wave formation, starting with beginning of effective stroke. (After Jahn and Bovee, 1967.)

up with a planar, distally propagating waveform all along its longer flagellum (Lowndes, 1936, 1941; Chen, 1950; Holwill, 1966b), which appears to increase in wavelength and amplitude distally (Lowndes, 1941) and is generally irregular. (3) When *P. trichophorum* swims freely forward, flagellum-first, a three-dimensional waveform passes toward the base along its anterior flagellum (Chen, 1950; Jahn et al., 1964a; Holwill, 1966b; Jahn and Bovee, 1967). The waveform has been described as helical (Jahn et al., 1964a; Jahn and Bovee, 1967); however, Chen (1950) described the basal region as having a conical gyration and the tip as bending and undulating as during gliding.

Lowndes (1941) reported that *Menoidium incurvum* swam at a speed of 50 μm s^{-1}. Its emergent flagellum was about 25 μm long. It beat at a rate of about 12 s^{-1} and had only one bend at a time traveling from base to tip.

Lowndes (1941) also reported observations of *Distigma*. One flagellum was over four times the length of the other. When freely swimming, the longer one bent back and passed bends from base to tip, in a manner resembling that of *Euglena,* and on rare occasion the shorter one swung forward. When creeping on a slide the flagellum was directed anteriorly.

A ciliumlike beat at the tip has been reported in the emergent flagellum of *Petalomonas* and the anterior flagellum of *Entosiphon sulcatum* (Jahn et al., 1963a; Jahn and Bovee, 1967). In *Petalomonas* the beating involves only about the distal fourth of the flagellum, whereas in *E. sulcatum* it involves about 90 percent of the flagellum.

Chlorophyceae

The waveforms of *Chlamydomonas* flagella can be considered the "type species" of green algal waveforms. They are by far the best-studied algal waveforms and where waveforms of other green algae have been described, their descriptions are usually consistent with those of *Chlamydomonas*. *Chlamydomonas* cells typically swim in a helical path. Although Ringo (1967) and Racey et al. (1981) did not find them to rotate, numerous workers have noted that they rotate about their longitudinal axis as they swim forward (e.g., Ulehla, 1911; Boscov and Feinleib, 1979; Foster and Smyth, 1980; Nultsch, 1983; Kamiya and Witman, 1984). They rotate counterclockwise as seen from behind when swimming freely (Rüffer and Nultsch, 1985) or when swimming in circles against a slide (Kamiya and Witman, 1984), which implies that they swim in left-handed helical paths. Kamiya and Witman (1984) found that their strain of *C. reinhardtii* swam with its eyespot toward the outside of the helical path, whereas Rüffer and Nultsch (1987) found that their mutant strain of *C. reinhardtii* (622 E) swam with the eyespot toward the inside. They have two flagella that emerge anteriorly and beat with identical or very similar waveforms. The flagellum on the eyespot side of the cell is referred to as the *cis-flagellum;* the other one is called the *trans-flagellum.* The flagella have a complex surface structure and bear fine hairs, but they do not have mastigonemes (see Moestrup, 1982, for review). They exhibit two well-defined waveforms (e.g., Ringo, 1967). In the usual, "forward," swimming mode, each flagellum has a very asymmetric stroke resembling those of cilia (e.g., Sleigh, 1974), so that their beating is reminiscent of a swimmer's breast stroke, as shown in Figure 4.7a–e. They exhibit a "reverse" swimming mode, in which the flagella beat symmetrically, passing bends from base to tip and propelling the cell like a pair of sperm tails, as shown in Figure 4.7f. The highly asymmetric forward waveforms can be thought of as being similar to the symmetric reverse waveforms, except that in the forward waveform the bend whose concave side faces laterally ("principal bends") are very deep and the bends whose concave side faces medially ("reverse bends") are very shallow. The formation of the deep principal bends at the basal end causes the flagella to be directed outward and posteriorly during the effective stroke. The propagation of the principal bends to the tip produces the recovery stroke. The striated basal fibers are not necessary for switching between the forward and reverse modes (Hoops et al., 1984). A closely related alga, *Chlorogonium elongatum,* can switch without even basal bodies (Hoops and Witman, 1983a).

Racey et al. (1981) found that when beating in the forward mode, *C. reinhardtii* had an average forward speed of 379 μm s^{-1} during the effective stroke and an average backward speed of 218 μm s^{-1} during the recovery stroke, with an average net forward swim speed of 124 μm s^{-1}. Brokaw and Luck (1983) have noted a corresponding forward-and-backward rotational movement in rotating uniflagellate cells. Goodenough (1983) found live vegetative cells and gametes of *C. reinhardtii* to swim with forward speeds of 133 μm s^{-1} and 174 μm s^{-1}, respectively, whereas reactivated demembranated models of the same cells swam at about half these speeds in 1 mM ATP.

The gametes of *Chlamydomonas* closely resemble the vegetative cells in structure and flagellar waveforms. During mating, (+) and (−) gametes agglutinate by their flagella, which continue to beat actively (e.g., Sager and Granick, 1954). Gametes release a factor into the water that causes agglutination of cells of the opposite mating type (e.g., Moewus, 1933; Förster and Wiese, 1954, 1955), but there is no evidence of chemotactic behavior.

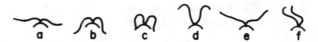

Figure 4.7. Waveforms of *Chlamydomonas reinhardtii.* (a–e) forward swimming, starting with end of effective stroke. (f) reverse swimming. (From unpublished film sequence by author.)

Brokaw et al. (1982) and Brokaw and Luck (1983) have analyzed the waveforms of *C. reinhardtii* in detail, using gametes of a mutant (*uni* − 1) that has only the trans-flagellum; it was easier to film than the wild type because it rotated about the cell body instead of swimming through the medium. Stroboscopically illuminated dark-field images made at 300 Hz on moving film were digitized for analysis. Usually about 60 points were recorded along a flagellum about 12 μm in length. These were combined to give flagellar angles at 0.4-μm intervals. They found that the waveforms could be approximated reasonably well by circular arcs and interconnecting straight regions (cf. Brokaw and Wright, 1963; see earlier discussion of chrysomonads). In normal-length flagella beating in the forward mode, principal bends had an angle of 3.68 rad and a curvature of 0.61 rad μm^{-1}, whereas reverse bends had an angle of only 0.88 rad and a curvature of only 0.09 rad μm^{-1}. In the reverse mode there was much less difference between principal and reverse bends: principal bends had an angle of 2.64 rad and a curvature of 0.43 rad μm^{-1}, and the corresponding values for the reverse bends were 2.66 rad and 0.39 rad μm^{-1}. Beat frequencies in the forward and reverse modes were 63.0 Hz and 73.3 Hz, respectively, at 25°C. A number of other beat parameters are presented in Brokaw and Luck (1983). Eshel and Brokaw (1987) have shown that the asymmetric forward waveform can be represented as a symmetric beating superimposed upon a flagellum of constant curvature. This means that the change between symmetric and asymmetric waveforms could be separate from the periodic sliding associated with the formation and propagation of bends.

The forward and reverse waveforms of *C. reinhardtii* appear at least approximately planar, with both flagella beating in the plane that contains the longitudinal cell axis (e.g., Ringo, 1967). However, Rüffer and Nultsch (1985, 1987), using high-speed films of a mutant (622 E) that swims more smoothly than wild type, found that there is a nonplanar component to the forward waveform. The tip turns out of the plane at the end of the recovery stroke; the nonplanar component moves proximally during the effective stroke.

The waveforms and beat frequencies of the two flagella usually appear to be at least approximately identical, and they often appear to beat synchronously. Hyams and Borisy (1975) found synchronous beating in flagellar apparatuses that had been isolated from a wall-less mutant of *C. reinhardtii* and reactivated in ATP, indicating that the synchrony was intrinsic to the flagellar apparatus. However, the synchrony is not perfect. Rüffer and Nultsch (1985, 1987) found that, on average, once in about every 20 beats the flagellum toward the outside of the helical swim path (the trans-flagellum) of mutant 622 E produced an additional beat. More than one rapid beat or, rarely, continuous asynchronous beating was also seen. They noted several other differences between the cis- and trans-flagellum. The recovery stroke of the cis-flagellum was closer to the cell axis (i.e., more in front of the cell). The bend developed earlier in the effective stroke in the cis-flagellum, so that the backward-moving portion of the flagellum was shorter in the cis-flagellum than in the trans-flagellum during the effective stroke. They also thought that the trans-flagellum beat farther out of the flagellar plane. They thought that the differences in waveform were more important than the beat frequency differences in generating rotation and helical motion. Kamiya and Hasegawa (1987), using fast Fourier transforms to measure beat frequencies, also found differences in beat frequency between the cis- and trans-flagellum. Live cells beat at 48–53 Hz at 20°C. Reactivated detergent-extracted cells exhibited two beat frequencies, about 30 Hz and 45 Hz. In reactivated cells with only one flagellum beating, the cis-flagellum had a lower beat frequency. When cells were mechanically rendered uniflagellate, the trans-flagellum beat with about a 30 percent higher frequency. In intact cells, both flagella beat at the frequency of the cis-flagellum. They did not report whether the two flagella beat in phase, as these flagella usually do. Changing the beat pattern of the trans-flagellum to synchronize it with the cis-flagellum over the entire beat cycle would, of course, require more than a change in the frequency of initiation of beats; it would also either produce a change in the waveform or require a change in the propagation speed of bends.

Chlamydomonas cells normally swim in the forward direction. When stimulated with a rapid increase in light intensity, they switch to the reverse mode for a second or two and then return to the forward mode. Schmidt and Eckert (1976), using a live wall-less mutant, showed that this "photophobic" response of *C. reinhardtii* depends on Ca^{2+} in the medium. In 10^{-3} M Ca^{2+} they showed the following response to a 280-ms pulse of light at 20°C: Cells continued to swim forward for about 20 ms after the onset of the stimulus, stopped for about 20 ms, then switched to a reverse waveform and beat actively for 200 ms after the stimulus. After 500 ± 100 ms, the cell slowed and wobbled with a reverse waveform for about 200 ms before resuming forward beating. Decreasing the Ca^{2+} concentration did not appreciably alter the forward waveform but inhibited reverse beating. At 10^{-6}–10^{-5} M Ca^{2+}, the reverse waveform appeared but beating was interrupted; below 10^{-6} M Ca^{2+}, no reverse beating was observed. These results suggested that reversal involved a transient increase in intracellular Ca^{2+} from the medium. These results are consistent with the findings of Hyams and Borisy (1975, 1978) that flagellar apparatuses isolated from a wall-less mutant of *C. reinhardtii* exhibited a forward waveform when reactivated in Ca^{2+} concentrations less than 10^{-7} M and a reverse waveform in Ca^{2+} concentrations greater than 10^{-7} M. The beat frequency also varies with Ca^{2+} concentration. Omoto and Brokaw (1985) found that frequency and asymmetry do not change in a parallel fashion as Ca^{2+} concentration is varied, suggesting that they may reflect different actions of Ca^{2+}. Bessen et al. (1980) found that when axonemes of *C. reinhardtii* isolated from the cell body with dibucaine and demembranated with detergent were reactivated in free Ca^{2+} concentrations of 10^{-6} M or less, they beat with a planar asymmetric waveform similar to that seen in forward swimming; when reactivated in 10^{-4} M free Ca^{2+}, they beat with a symmetric waveform similar to that seen in reverse swimming. In 10^{-5} M free Ca^{2+} they were predominantly quiescent. These results indicate that the Ca^{2+} acted directly on the axonemes. The photophobic action of Ca^{2+} may be mediated by a phosphoprotein (Segal et al., 1983, 1984; Segal and Luck, 1985).

In addition to having a photophobic response, *Chlamydomonas*, like many algal cells, exhibits phototaxis. Kamiya and Witman (1984), using demembranated cells of *C. reinhardtii*, found that the percentage of cells with active trans-flagella was less than the percentage of cells with active cis-flagella when reactivated in less than 10^{-8} M Ca^{2+}; the situation was reversed at higher concentrations. They suggested a model in which live cells use a calcium-mediated differential activity of the cis- and trans-flagella to make tactic turns toward or away from a light source. The data of Morel-Laurens (1987) are consistent with this model. Smyth and Berg (1982), using *uni*–1 mutants with only the trans-flagellum, found a decrease in the beat frequency of this flagellum in response to changes in light level. The decrease was more prolonged during a change characteristic of a stimulus for a negative phototactic response. They suggested that the cis-flagellum might show reciprocal responses, with a more prolonged decrease of beat frequency during positive phototaxis. However, the waveforms of the flagella of *Chlamydomonas* have not yet been analyzed during phototactic turns. Phototaxis in algae has been reviewed recently by Nultsch and Häder (1988).

Mutants of *Chlamydomonas* lacking particular axonemal structures have been used to provide clues about possible functions of the missing components. *C. reinhardtii* can beat without the central pair of tubules (Goldstein, 1982; Luck et al., 1982; Brokaw and Luck, 1985). They beat in the same plane as intact flagella and can beat in both the forward and reverse modes without a central pair. The direction of beating may be determined by the orientation of the outer doublets (Hoops and Witman, 1983b). They can also beat without intact radial spokes (Brokaw et al., 1982; Brokaw and Luck, 1985). Brokaw et al. (1982), using suppressor mutations to restore swimming to mutants lacking a radial-spoke system (*pf*–17), without restoring the missing radial-spoke structures, found their waveforms to be appreciably more symmetric than the wild type. They suggested that the radial-spoke system is required for increasing the efficiency of swimming by inhibiting the development of reverse bends during the recovery stroke. Hosokawa and Miki-Noumura (1987) obtained similar results on wild-type *C. reinhardtii* in which the central pair was extruded from demembranated flagella with elastase digestion. They obtained symmetric beating in concentrations of

Ca^{2+} between 10^{-9} and 10^{-4} M and did not see the asymmetric waveform. A variety of mutants that lack inner and/or outer dynein arms have been reported (e.g., Goodenough and Heuser, 1985; Kamiya and Okamoto, 1985; Mitchell and Rosenbaum, 1985; Okagaki and Kamiya, 1986; Brokaw and Kamiya, 1987; Sakakibara and Kamiya, 1989). They exhibit a variety of changes from wild-type beating; a lack of outer arms appears to be associated with decreased bend amplitude and a decreased ability to form a reverse waveform.

The biflagellate cell *Polytoma uvella* swims with a speed of about 92 μm s^{-1} at 20°C (Gittleson and Noble, 1973). Brokaw (1963), using stroboscopic illumination and stroboscopically illuminated multiple-exposure plates, found that the forward waveforms of the flagella of *P. uvella* resembled those of *Chlamydomonas*. They were about 10–25 μm long and beat at 10–20 Hz in growth medium at 14°C, and at higher frequencies at higher temperatures and when the Ca^{2+} was removed. Under proper conditions the beat frequency averaged about 50 Hz. However, there was usually little correlation between the beat frequencies of the two flagella, and they could both be beating apparently normally with the frequency of one being 50 percent greater than that of the other. In fact, one flagellum could be beating normally while the other one was quiescent. This difference may have been responsible for the asymmetry that produced their helical swim paths. When the viscosity of the medium was increased with methyl cellulose, the waveforms did not change noticeably even when the viscosity was great enough to reduce the beat frequency to about 2 Hz. However, the waveforms became quite symmetric when the viscosity was great enough to slow the beat frequency to 1–2 Hz. Even at this viscosity the beat amplitude remained constant along the flagellum instead of diminishing distally. When Ca^{2+} was present, the cells exhibited frequent "shock reactions," in which they darted backwards for 2–3 s. During this response the flagella were thrown forward; their waveform was symmetric and resembled the reverse waveform of *Chlamydomonas*. The flagella could be removed from the cell bodies and reactivated in solutions containing ATP after their membranes were disrupted with glycerol (Brokaw, 1961, 1962, 1963). The reactivated flagella could beat at frequencies up to 20 Hz, at 10 mM ATP, and propagated bends distally. They beat with a fairly planar waveform, which deviated from planar somewhat in their distal half. When they were freely swimming, their waveform was rather symmetric, but it became noticeably asymmetric when the flagellar base was attached to the glass.

Gittleson and Jahn (1968) found the flagella of the quadriflagellate cells of *Polytomella agilis* to be 8.0–9.0 μm in length. They arose anteriorly and had a planar waveform similar to the forward waveform of *Chlamydomonas*. However, they were unusual in that the effective stroke lasted longer (70–90 percent of the beat cycle) than the recovery stroke. They beat in pairs, with one pair on each side of the cell. The two flagella of each pair beat synchronously with one another. The degree of coordination between the two pairs of flagella is not completely clear. Films of swimming cells showed the two pairs to beat at appreciably different frequencies. However, this lack of coordination may have been at least partly due to the intense illumination used for filming; the authors state that there was "a breast stroke-like coordination" between the flagella on the two sides of the cell. They swim at a speed of about 233 μm s^{-1} (Gittleson and Noble, 1973).

Watson (1975) found the 10-μm-long flagella of the biflagellate zoospores of *Microthamnion kuetzingianum* to beat with distally directed flagellar undulations during both forward and reverse swimming. The flagella appear to be about one wavelength long in the photographs. During forward swimming, the flagella could be either in phase or about 180° out of phase with one another. Under normal swimming conditions, about 85 percent of the cells swam forward, 4 percent swam backward, and 11 percent were quiescent, with their bases directed about 40°–70° from the horizontal, at any time. The cells switched from forward to reverse swimming by changing the angle between the basal bodies. In forward-swimming cells, both basal bodies were directed anteriorly, whereas in reverse-swimming cells the basal bodies were oriented latero-posteriorly. Cells whose flagella had been sheared off were still capable of reorienting the basal stubs, indicating that this ability does not reside in the flagellar axoneme.

Spermatozopsis similis swims with waveforms similar to those of *Chlamydomonas*, with an asymmetric breast stroke during forward swimming and a symmetric, anteriorly directed flagellar waveform during reverse swimming (Preisig and Melkonian, 1984; Melkonian and Preisig, 1984; McFadden et al., 1987). The flagella are somewhat unequal in length, the longer one being 15–20 μm long and the shorter one 8–16.5 μm long. However, although *Chlamydomonas* maintains a reasonably constant angle between its basal bodies during reversal, the angle between the basal bodies of *S. similis* changes in a manner similar to that seen in *M. kuetzingianum*. During forward swimming, the basal bodies are oriented more or less antiparallel to one another, so that the basal ends of the flagella project laterally; during reversal the basal bodies are parallel, so that the basal ends of the flagella project anteriorly. McFadden et al. (1987), using reactivated flagellar apparatuses, obtained both a transition from asymmetric to symmetric waveforms and a transition from antiparallel to parallel orientation of the basal bodies when increasing the Ca^{2+} concentration from about 10^{-8} to 10^{-6} M. This change in orientation of the basal bodies was accompanied by a shortening of the distal connecting fiber between them. The results on *S. similis* and *M. kuetzingianum* indicate that basal body reorientation in these cells is due to Ca^{2+}-mediated basal contractions independent of changes in flagellar beating.

Friedmann (1959) found the male, but not the female, gametes of *Praisiola stipitata* to be flagellate. They had two flagella, inserted anteriorly, "either facing each other diametrically or at an obtuse angle." The flagella were smooth and of equal length, 4–14 µm long. They swam "with a peculiar rocking movement which could be described as 'paddling' with their flagella." That is, "it swims, as do all isokont chlorophycean swarmers, with its flagella directed forwards." When it reaches an egg, it "gives the impression it 'feels' the egg by its flagella. Then it starts to circle around an egg, gliding on the surface of the latter and in constant contact with it by means of one of its flagella." The gametes then fuse as the flagellum in contact with the egg becomes absorbed into the egg's cytoplasm. The two flagella are rigid and diametrically opposed at this point. The pair of fusing gametes "swim with the egg pointing forwards and the free, unfused flagellum backwards. Both the egg and the tip of the free flagellum revolve in opposite directions and hence the 'axis' formed by the two flagella describes a surface of two cones of revolution joined by their apices." When fusion is complete, "the pear-shaped zygote swims vigorously, its flagellum pointing backwards and the whole cell revolving around its long axis. . . ." The cell is thus transformed from an anteriorly biflagellate gamete into a posteriorly uniflagellate zygote. It would be interesting to know how the flagellar waveform changes during this transformation.

Both flagella of the spermatozoids of *Golenkinia minutissima* have a "9 + 1" ultrastructure, with one central microtubule instead of a central pair (Moestrup, 1972). The central microtubule has a sinuous appearance and may not be functional. The outer doublets appear normal. Although no detailed analysis has been made of the motility of these flagella, they appear to beat in a manner very similar to that of *Chlamydomonas*.

The colonial Volvocales have vegetative and reproductive cells that resemble *Chlamydomonas*. *Gonium pectorale* is a more or less planar colony of 16 cells, which swims with its plane perpendicular to the direction of movement. Moore and Goodspeed (1911) and Moore (1916) described the motility of its flagella, using observations on isolated cells. Occasionally the flagella were motile over their entire length, but generally only the distal half or third beat while the proximal portion appeared practically rigid. Although they reported that the distal portion "describes a cone, the effective component of which is backward," their techniques may not have allowed them to distinguish between the described waveform and a ciliumlike beat. They noted that cells could turn by (1) keeping one flagellum inactive while the other one beat normally or (2) having both flagella produce an effective stroke in the same direction. Both of these types of behavior are quite different than those of *Chlamydomonas*.

The colonies of *Volvox* are spherical or oval. These colonies do not reverse their swim direction, but can turn (e.g., Mast, 1926). When *V. carteri* (e.g.,

Sakaguchi and Iwasa, 1979) and V. *aureus* (e.g., Hand and Haupt, 1971; Sakaguchi and Tawada, 1977) are illuminated from the side at room temperature, they turn toward the light by stopping or slowing the anterior flagella on the side of the colony nearer the light. At lower temperatures the phototaxis becomes negative. The waveform of the flagella of *Volvox* has not been studied in detail.

The Tetrasporales can have biflagellate vegetative cells and reproductive cells. However, the vegetative cells are only weakly motile in the Palmellaceae (e.g., King, 1973), and the zygotes give rise to immotile cells in the Tetrasporaceae (e.g., Klyver, 1929). In the Tetrasporaceae the vegetative cells have long immobile "pseudocilia." Lembi and Herndon (1966) and Wukej (1968) have examined the ultrastructure of the pseudocilia in *Paulschulzia pseudovolvox*, *Tetraspora gelatinosa*, and *Schizochlamys gelatinosa*. The pseudocilia of *P. pseudovolvox* were usually 75–100 μm long, but could be up to 200 μm. In all three species they had a "9 + 0" cross section for a short distance near their base. They became reduced to four or five singlets for most of their length, and in *P. pseudovolvox* they could have as few as two singlets at the tip. In *S. gelatinosa*, "many cells bear pseudocilia that fork . . . , so that each cell apparently bears two to eight pseudocilia."

Flagellate reproductive cells of the Oedogoniales are "stephanokonts"; that is, they have a ring of flagella at their anterior end. In *Oedogonium cardiacum* the zoospores have a ring of about 120 flagella (Hoffman and Manton, 1962), and the spermatozoids have a ring of about 30 flagella (Hoffman and Manton, 1963). In the zoospores the average flagellar length is 17 μm; in the spermatozoids it is 16.5 μm. The gametes exhibit chemotaxis toward the oogonia (Hoffman, 1973) and toward extracts of oogonia and female filaments (Hoffman, 1960). The flagellar waveforms of *Oedogonium* have not been analyzed.

Ulvophyceae

The Ulvophytes do not have flagellate vegetative cells, but they usually have biflagellate gametes and/or bi- or quadriflagellate zoospores [although Prasad and Srivastava (1963) have reported getting swarmers of *Schizomeris leibleinii* with two to eight or more flagella]. The gametes commonly exhibit positive phototaxis until they fuse, when they quickly switch to negative phototaxis (e.g., Bliding, 1957; Chihara, 1969; Bråten, 1971; Melkonian, 1980).

Although the Derbesiaceae have biflagellate gametes, they produce stephanokont zoospores; that is, the zoospores have a ring containing many flagella (e.g., Rietema, 1969, 1971). The male and female gametes of *Bryopsis hypnoides* have apically inserting, equal, smooth flagella 20–40 μm long (Burr and West, 1970). The gametes become motile after exposure to light but before their explosive release from the gametangium. The males have much greater swim speeds than the females but generally stop swimming much sooner.

The flagella of the biflagellate gametes of *Dichotomosiphon* (Moestrup and Hoffman, 1975) and *Acetabularia* (Crawley, 1966) emerge from the cell apex and extend away from one another, at somewhat less than 180°.

Flagellar waveforms in the Ulvophyceae have not been analyzed.

Charophyceae

Vegetative cells of the stoneworts are not motile. However, reproductive cells in all orders but the Zygnematales have two flagella.

The Charales have elongate spermatozoids that resemble some animal sperm more than typical flagellate algal cells. *Nitella missouriensis* sperm are about 60 μm long (Turner, 1968). The flagella are about the same length and are attached subterminally at the anterior end and are directed posteriorly. The sperm can be seen moving within the antheridium. The sperm of *Chara fibrosa* (Pickett-Heaps, 1968) and *C. corallina* (Moestrup, 1970b) have a similar form. The flagella of *C. fibrosa* were found to have different lengths, the longer one being 56.5–68.3 μm long and the shorter one 49.3–60.6 μm long. The flagella could be seen beating vigorously within the mother cell after the rest of the sperm had emerged.

The Klebsormidiales and the Coleochaetales produce zoospores. In *Klebsormidium flaccidum* (Marchant et al., 1973), *Coleochate pulvinata* (Sluiman, 1983), and *C. scutata* (Pickett-Heaps and Marchant, 1972), both flagella project laterally from the cell body. The zoospores of *C. scutata* swim with a circular motion for a few minutes after release and do not get very far from the parental thalli (Graham et al., 1986).

The flagellar waveforms of the Charophyceae await analysis.

Prasinophyceae

The number of flagella varies with species in the prasinophytes, being one or an even number from two to eight. The flagella generally bear scales and fine hairlike projections called "hair-scales" (e.g., Norris, 1980; Moestrup, 1982). It has been suggested that the hair-scales, which appear to be anchored to outer doublets, may change their angle relative to the flagellar axis and amplify the effects of beating (Melkonian, 1982). In some species the flagella push the cell from behind, rather than pulling from the anterior end as they do in most algae. In uniflagellate or biflagellate cells the flagella insert laterally; in the others the flagella insert apically. The inclusion of such disparate cells in a single class (e.g., Norris, 1980) has been opposed by some workers (e.g., Stewart and Mattox, 1975).

The single flagellum of *Micromonas pusilla* emerges from the concave side of the comma-shaped cell body and curves back, so that the cell is pushed from behind (e.g., Manton, 1959b; Manton and Parke, 1960; Omoto and Witman, 1981). The cell swims about 90 μm s^{-1} (Throndsen, 1973) and rotates when swimming. In many algal flagella the central pair of microtubules extends somewhat beyond the outer doublets, forming a thin projection at the tip (a "hairpoint"). However, in *M. pusilla* this projection is extreme. The full 9 + 2 cross section is less than 1 μm long, while the "0 + 2" projection is about 4 μm long. Omoto and Witman (1981) have analyzed the motility of this flagellum. When swimming, the 0 + 2 portion formed a left-handed helix and rotated clockwise as viewed from in front the cell, pushing the cell from behind. The cell body in turn rotated counterclockwise. They interpreted the rotation of the 0 + 2 portion as indicating a clockwise rotation of the central pair within. The 9 + 2 basal portion appeared to wave back and forth. The dimensions and frequency of rotation of the helix were not given, but the flagella in their figures are about one turn long, and the flagellum in their Fig. 2 appears to be rotating about six or seven turns per second. The 0 + 2 portion retained its helical form in nonmoving cells. *M. pusilla* is the only cell known that appears to use rotation of the exposed central pair to develop thrust. Manton and Parke (1960) reported that *M. pusilla* could also swim for short periods with the flagellum directed forward, but that the speed was slower than when swimming with the flagellum trailing. The waveform for this mode of swimming was not described.

Manton and Parke (1960) found that *Pedinomonas tuberculata* swam with its single flagellum trailing from its insertion at the posterior end of the cell body. At rest or when the cell changed swimming direction, the flagellum curved toward the cell body, so that it pointed toward the anterior end of the cell. Ettl (1964) recorded a similar behavior in *P. minor*, but from a different perspective; he considered the flagellum to emerge from the anterior end. He reported that the flagellum curved over the anterior end when resting and that during swimming it was extended, making the cell swim with its posterior end leading. Swimming cells rotated rapidly and vibrated somewhat. Kalinsky (1971) also found that the flagellum of *P. minor*, which was 6.8–7.0 μm long, trailed when swimming and was compressed along the hyaline region of the cell when resting. Sweeney (1971, 1976) noted that the flagellum of *P. noctilucae*, a symbiote of the dinoflagellate *Noctiluca miliaris*, was posteriorly inserted and directed toward the rear during swimming. Curiously, although *P. noctilucae* was uniflagellate within the vacuole of *N. miliaris*, when it was cultured in enriched seawater media the resulting cells "possessed 2 anterior flagella and resembled *Chlamydomonas*." A hint of a second basal body was seen in the uniflagellate cells with electron microscopy, but it was not clearly defined. The waveforms of *Pedinomonas* have not been analyzed.

Barlow and Cattolico (1980) found that *Mantoniella squamata* had two flagella. Both flagella bore some fine hairs as well as scales. The longer one was 2.5–4.0 times its 2.4–4.0 μm body length, but the other one was 1.0 μm long or less. They reported that the longer flagellum "emerges laterally but is directed behind the cell during swimming." However, Manton and Parke (1960) reported that the longer flagellum (which they took to be the only flagellum) bent back sharply from a 1–2 μm long, spinelike projection at the anterior end of the cell body. Flagellar undulations, usually small but occasionally large, could be seen. The cell stopped swimming by rapidly curving the flagellum around the cell body. It could be coiled for two turns around the cell. It then uncoiled the flagellum and swam off in a new direction. Before cell division, the two long flagella beat homodynamically, and even twisted around the cell body simultaneously.

Nephroselmis has two flagella that emerge laterally (e.g., Moestrup and Ettl, 1979; Moestrup, 1983; Melkonian et al., 1987). In *N. olivacea* the length of one flagellum (18–25 μm) is usually at least one and one-half times the length of the other (12–16 μm). When swimming, the short flagellum is bent anteriorly and the longer one is bent posteriorly. The basal ends of the flagella enter the side of the cell as it swims. Moestrup and Ettl (1979) found that the flagella moved in a plane and that the shorter flagellum of *N. olivacea* generated the forward thrust whereas the longer one trailed passively. Melkonian et al. (1987) found that the shorter one beat "in a cilia-type mode," whereas the longer one trailed and exhibited "undulating waves passing from base to tip." Neither waveform has been analyzed in detail.

Tetraselmis has a zig-zag row of four flagella that emerge anteriorly (e.g., Melkonian, 1979; Salisbury et al., 1981). Salisbury et al. (1981) found that the flagella of *T. subcordiformis* beat in a ciliary fashion, with an oarlike effective stroke followed by a return stroke, similar to the beat of *Chlamydomonas*. Normally, the beating of all four flagella was coordinated, and during forward swimming the flagella of each outer pair beat together "in a nearly planar fashion." Occasionally, however, alternate flagella would beat slowly in opposite directions. Resting cells were often seen with all four flagella curved back along the sides of the cell, in a position corresponding to the end of the effective stroke. Salisbury and Floyd (1978) described a contractile rootlet structure (a "rhizoplast") in *T. subcordiformis,* which contracted rhythmically when the cells were placed in artificial seawater containing 2 mM $CaCl_2$ and 5 mM ATP and displaced the basal bodies inward from the anterior end. (See chapter 6 for more details on centrin-mediated cell motility). They suggested that this displacement could be involved in the control of flagellar coordination or beat direction. The possibility is an interesting one. The exact mechanism suggested, of inhibiting basal *Chlamydomonas*-like bending by pulling the flagella into the apical pit, seems unlikely because *Chlamydomonas* does not exhibit the extreme basal flexion they assumed.

However, the flagellar waveform *T. subcordiformis* has not yet been analyzed in detail and may have more extreme basal bends than *Chlamydomonas*.

Scherffelia dubia has four anteriorly inserted flagella of equal length (8–12 μm), which emerge in two pairs (Melkonian and Preisig, 1986). When at rest the pairs are curved back along opposite sides of the cell body, with the two flagella of each pair lying near one another.

Depending on the species, *Pyramimonas* has either four or eight flagella, which emerge from a depression at the anterior end of the cell body. The flagella are covered with scales (Melkonian, 1981). The flagella beat with a ciliumlike stroke (e.g., Belcher, 1969; Throndsen, 1973; Hori and Moestrup, 1987; Moestrup et al., 1987) similar to that found in *Chlamydomonas* and can exhibit a similar reversal response (e.g., Belcher, 1969). Throndsen (1973) found *P.* cf. *disomata* to swim rapidly in a helical path, making 4–6 gyres s^{-1} and to have an average swim speed of 350 μm s^{-1}. The diameter of the path decreased with increasing swimming speed, to two to three cell diameters at 250 μm s^{-1}. *P. tetrarhynchus* has four flagella. Belcher (1969) found them to swim about a millimeter in 8 s (i.e., about 125 μm s^{-1}). Observed with stroboscopic dark-field illumination, the flagella were found to be directed obliquely backward during their entire beat cycle. Waves passed from base to tip, with the wavelength being about one half to two thirds the length of the flagellum. The beating was planar. The beat plane was at a slight angle to the longitudinal axis of the cell; this angle was thought to cause the rotation seen during swimming. When a cell changed direction, the flagella were suddenly directed forward; the cell stopped instantaneously and swam off immediately in another direction. *P. octopus* has eight flagella (sixteen when dividing). Hori and Moestrup (1987) measured the speeds of two cells of *P. octopus* as 214 μm s^{-1} and 241 μm s^{-1}. The cells swam until they attached to sand, other cells, or the slide. Flagella of attached cells vibrated slightly, having "very rapid movements of small amplitude." When disturbed, the cells swam against the slide in circles (90–125 μm in diameter in the two cells measured), apparently always clockwise as seen from above. They analyzed the orientation of the effective strokes of the eight flagella, as surmised from the orientation of the outer doublets. When the cells were at rest, the flagella spread out evenly, about 45° apart. However, during swimming the effective strokes appeared not to be entirely radial. They concluded that the angles of the effective strokes would cause swimming cells to rotate in the observed direction. Moestrup et al. (1987) found that the flagella emerged apically, curved back, and extended beyond the back of the cell. They could be almost straight distally, but more often the distal portion curved away from the cell body.

Conclusion

Although the basic flagellar configuration in algal cells consists only of two emergent flagella, a large array of variations on this simple theme has allowed a remarkable diversity of patterns of flagellar beating. The more the motility of these cells has been studied, the more complex their behavior has proven to be. The descriptions of the waveforms in a number of groups of algae are quite scanty. Even so, the observations that have been reported for them suggest that there are exotic movements yet to be seen, and perhaps equally exotic mechanisms yet to be discovered.

References

Afzelius, B. A. 1962. The contractile apparatus in some invertebrate muscles and spermatozoa. In S. S. Brese ed., *Fifth Int. Cong. Electron Microsc.* 2:M–1. Academic Press, New York.

Afzelius, B. A. 1969. Ultrastructure of cilia and flagella. In Lima-de-Faria ed., *Handbook of Molecular Cytology.* North Holland, Amsterdam, pp. 1219–1241.

Andersen, R. A. 1987. Synurophyceae classis nov., a new class of algae. *Am. J. Bot.* 74:337–353.

Baccetti, B., Burrini, A. G., Dallai, R., and Pallini, V. 1979. The dynein electrophoretic bands in axonemes naturally lacking the inner or the outer arm. *J. Cell Biol.* 80:334–340.

Barlow, S. B., and Cattolico, R. A. 1980. Fine structure of the scale-covered green flagellate *Matoniella squamata* (Manton & Parke) Desikachary. *Br. phycol. J.* 15:321–333.

Belcher, J. H. 1969. Further observations on the type species of *Pyramimonas* (*P. tetra-hynchus* Schmarda) (Prasinophyceae): An examination by light microscopy, together with notes on its taxonomy. *Bot. J. Linn. Soc.* 62:241–253.

Berdach, J. T. 1977. In situ preservation of the transverse flagellum of *Peridinium cinctum* (Dinophyceae) for scanning electron microscopy. *J. Phycol.* 13:243–251.

Bessen, M., Fay, R. B., and Witman, G. B. 1980. Calcium control of waveform in isolated flagellar axonemes of *Chlamydomonas. J. Cell Biol.* 86:446–455.

Billard, C. 1983. *Prymnesium zebrinum* sp. nov. et *P. annuliferum* sp. nov., deux nouvelles espèces apparentées à *P. parvum* Carter (Prymnesiophyceae). *Phycologia* 22:141–151.

Bliding, C. 1957. Studies on *Rhizoclonium* I. Life history of two species. *Bot. Not.* 110:271–275.

Boscov, J. S., and Feinleib, M. E. 1979. Phototactic response of *Chlamydomonas* to flashes of light. II. Response of individual cells. *Photochem. Photobiol.* 30:499–505.

Bouck, G. B. 1969. Extracellular microtubules: The origin, structure and attachment of flagellar hairs in *Fucus* and *Ascophyllum* antherozoids. *J. Cell Biol.* 40:446–460.

Bouck, G. B. 1971. The structure, origin, isolation and composition of the tubular mastigonemes of the *Ochromonas* flagellum. *J. Cell Biol.* 50:362–384.

Bouck, G. B., and Green, P. M. 1976. Paracrystals and mastigonemes are directly attached to the *Euglena* axonemal microtubules. *J. Cell Biol.* 70:156a.

Bouck, G. B., Rogalski, A., and Valaitis, A. 1978. Surface organization and composition of *Euglena*. II. Flagellar mastigonemes. *J. Cell Biol.* 77:805–826.

Bradley, D. E. 1966. Observations on some chrysomonads from Scotland. *J. Protozool.* 13:143–154.

Bråten, T. 1971. The ultrastructure of fertilization and zygote formation in the green alga *Ulva mutabilis* Tøyn. *J. Cell Sci.* 9:621–635.

Brennen, C. 1976. Locomotion of flagellates with mastigonemes. *J. Mechanochem. Cell Motil.* 3:207–217.

Brokaw, C. J. 1957. Electro-chemical orientation of bracken spermatozoids. *Nature* 179:525.

Brokaw, C. J. 1958a. Chemotaxis of bracken spermatozoids. The role of bimalate ions. *J. Exp. Biol.* 35:192–196.

Brokaw, C. J. 1958b. Chemotaxis of bracken spermatozoids: Implications of electrochemical orientation. *J. Exp. Biol.* 35:197–212.

Brokaw, C. J. 1959. Random and oriented movements of bracken spermatozoids. *J. Cell Comp. Physiol.* 54:95–101.

Brokaw, C. J. 1961. Movement and nucleoside polyphosphatase activity of isolated flagella from *Polytoma uvella*. *Exp. Cell Res.* 22:151–162.

Brokaw, C. J. 1962. Studies on isolated flagella. In D. W. Bishop (ed), *Spermatazoon Motility*. American Association for the Advancement of Science, Washington, D.C., pp. 269–278.

Brokaw, C. J. 1963. Movement of the flagella of *Polytoma uvella*. *J. Exp. Biol.* 40:149–156.

Brokaw, C. J. 1965. Non-sinusoidal bending waves of sperm flagella. *J. Exp. Biol.* 43:155–169.

Brokaw, C. J. 1968. Mechanisms of sperm movement. In P. L. Miller (ed.), *Aspects of Cell Motility (Symp. Soc. Exp. Biol. XXII)*. Cambridge University Press, London, pp. 101–116.

Brokaw, C. J. 1983. The constant curvature model for flagellar bending patterns. *J. Submicrosc. Cytol.* 15:5–8.

Brokaw, C. J. 1989. Direct measurements of sliding between outer doublet microtubules in swimming sperm flagella. *Science* 243:1593–1596.

Brokaw, C. J., and Kamiya, R. 1987. Bending patterns of *Chlamydomonas* flagella: IV. Mutants with defects in inner and outer dynein arms indicate differences in dynein arm function. *Cell Motil. Cytoskel.* 8:68–75.

Brokaw, C. J., and Luck, D. J. L. 1983. Bending patterns of *Chlamydomonas* flagella. I. Wild-type bending patterns. *Cell Motil.* 3:131–150.

Brokaw, C. J., and Luck, D. J. L. 1985. Bending patterns of *Chlamydomonas* flagella. III. A radial spoke head deficient mutant and a central pair deficient mutant. *Cell Motil.* 5:195–208.

Brokaw, C. J., Luck, D. J. L., and Huang, B. 1982. Analysis of the movement of *Chlamydomonas* flagella: The function of the radial-spoke system is revealed by comparison of wild-type and mutant flagella. *J. Cell Biol.* 92:722–732.

Brokaw, C. J., and Wright, L. 1963. Bending waves of the posterior flagellum of *Ceratium*. *Science* 142:1169–1170.

Brumpt, E., and Lavier, G. 1924. Un nouvel Euglénien polyflagellé parasite du têtard de *Leptodactylus ocellatus* au Brésil. *Ann. Parasitol.* 2:248–252.

Burr, F. A., and West, J. A. 1970. Light and electron microscope observations on the vegetative and reproductive structures of *Bryopsis hypnoides*. *Phycologia* 9:17–37.

Chadefaud, M., and Provasoli, L. 1939. Une nouvelle Euglène graciloïde: *Euglena gracilis* Klebs var. *urophora* n. var. *Arch. Zool. Exp. Gen. Notes Rev.* 80:55–60.

Chen, Y. T. 1950. Investigations of the biology of *Peranema trichophorum* (Euglenineae). *Quart. J. Microsc. Sci.* 91:279–308.

Chihara, M. 1969. Culture study of *Chlorochytrium inclusum* from the northeast Pacific. *Phycologia* 8:127–133.

Cook, A. H., and Elvidge, J. A. 1951. Fertilization in the Fucaceae: Investigations on the nature of the chemotactic substance produced by eggs of *Fucus serratus* and *F. vesiculosus*. *Proc. Roy. Soc.* B 138:97–114.

Cook, A. H., Elvidge, J. A., and Heilbron, I. 1948. Fertilization, including chemotactic phenomena in the Fucaceae. *Proc. Roy. Soc.* B 135:293–301.

Costello, D. P., Henley, C., and Ault, C. R. 1969. Microtubules in spermatozoa of *Childia* (Tubellaria, Acoela) revealed by negative staining. *Science* 163:678–679.

Crawley, J. C. W. 1966. Some observations on the fine structure of the gametes and zygotes of *Acetabularia*. *Planta* 69:365–376.

Darling, R. B., Friedmann, E. I., and Broady, P. A. 1987. *Heterococcus endolithicus* sp. nov. (Xanthophyceae) and other terrestrial *Heterococcus* species from antarctica: Morphologial changes during life history and response to temperature. *J. Phycol.* 23:598–607.

Desportes, I. 1966. L'ultrastructure du gamete male de l'Eugregarine *Stylocephalus longicollis*. *C. R. Acad. Sci.* (D) 263:517–520.

Dodge, J. D., and Bibby, B. T. 1973. The Prorocentrales (Dinophyceae) I. A comparative account of fine structure in the genera *Prorocentrum* and *Exuviaella*. *Bot J. Linn. Soc.* 67:175–187.

Dogadina, T. V. 1986. Groundlessness of distinguishing the alga group Eustigmatophyta. *Bot. Zh. (Leningr.)* 71:508–513.

Eppley, R. W., Holm-Hansen, O., and Strickland, J. D. H. 1968. Some observations on the vertical migration of dinoflagellates. *J. Phycol.* 4:333–340.

Eshel, D., and Brokaw, C. J. 1987. New evidence for a "biased baseline" mechanism for calcium-regulated asymmetry of flagellar bending. *Cell Motil. Cytoskel.* 7:160–168.

Ettl, H. 1964. Die feinere Struktur von *Pedinomonas minor* Korschikoff I. Untersuchungen mit dem Lichtmikroskop. *Nova Hedw.* 8:421–440.

Ettl, H. 1978. Xanthophyceae. In H. Ettl, J. Gerloff, and H. Henig (eds.), *Süsswasserflora von Mitteleuropa*. Gustav Fischer, Stuttgart, xiv + 530 pp.

Förster, H., and Wiese, L. 1954. Gamonwirkungen bei *Chlamydomonas eugometos*. *Z. Naturforsch.* 9b:548–550.

Förster, H., and Wiese, L. 1955. Gamonwirkungen bei *Chlamydomonas reinhardi*. *Z. Naturforsch.* 10b:91–92.

Forward, R. B., Jr. 1974. Phototaxis by the dinoflagellate *Gymnodinium splendens* Lebour. *J. Protozool.* 21:312–315.

Forward, R. B., Jr., and Davenport, D. 1970. The circadian rhythm of a behavioral photoresponse in the dinoflagellate *Gyrodinium dorsum*. *Planta (Berl.)* 92:259–266.

Foster, K. W., and Smyth, R. D. 1980. Light antennas in phototactic algae. *Microbiol. Rev.* 44:572–630.

Fresnel, J. 1986. Nouvelles observations sur une coccolithacée rare: *Cruciplacolithus neohelis* (McIntyre et Bé) Rheinhardt (Prymnesiophyceae). *Protistologica* 22:193–204.

Friedmann, I. 1959. Gametes, fertilization and zygote development in *Prasiola stipitata* Suhr. I. Light microscopy. *Nova Hedw.* 1:333–344.

Fuchs, B., and Jarosch, R. 1974. Rotierende Fabrillen in der *Synura*-Geissel. *Photoplasma* 79:215–223.

Gaines, G., and Taylor, F. J. R. 1985. Form and function of the dinoflagellate transverse flagellum. *J. Protozool.* 32:290–296.

Gayral, P., and Fresnel, J. 1979. *Exanthemachrysis gayraliae* Lepailleur (Prymnesiophyceae, Pavlovales): Ultrastructure et discussion taxinomique. *Protistologica* 15:271–282.

Gayral, P., and Fresnel, J. 1983. *Platychrysis pienaarii* sp. nov. et *P. simplex* sp. nov. (Prymnesiophyceae): Description et ultrastructure. *Phycologia* 22:29–45.

Geller, A., and Müller, D. G. 1981. Analysis of the flagellar beat pattern of male *Ectocarpus siliculosus* gametes (Phaeophyta) in relation to chemotactic stimulation by female cells. *J. Exp. Biol.* 92:53–66.

Gibbons, B. H., Baccetti, B., and Gibbons, I. R. 1985. Live and reactivated motility in the 9 + 0 flagellum of *Anguilla* sperm. *Cell Motil.* 5:333–350.

Gibbons, I. R. 1961. The relationship between the fine structure and direction of beat in the gill cilia of a lamellibranch mollusc. *J. Biophys. Biochem. Cytol.* 11:179–205.

Gittleson, S. M., and Jahn, T. L. 1968. Flagellar activity of *Polytomella agilis*. *Trans. Am. Microsc. Soc.* 87:464–471.

Gittleson, S. M., and Noble, R. M. 1973. Locomotion in *Polytomella agilis* and *Polytoma uvella*. *Trans. Am. Microsc. Soc.* 92:122–128.

Gojdics, M. 1953. *The Genus Euglena*. University of Wisconsin Press, Madison, 268 pp.

Goldstein, S. F. 1969. Irradiation of sperm tails by laser microbeam. *J. Exp. Biol.* 51:431–441.

Goldstein, S. F. 1982. Motility of 9 + 0 mutants of *Chlamydomonas reinhardtii*. *Cell Motil. Suppl.* 1:165–168.

Goldstein, S. F. 1983. Light microscopic analysis of flagellar waveforms: Some implications and current problems. *J. Submicrosc. Cytol.* 15:9–13.

Goldstein, S. F., and Schrével, J. 1982. Motility of the 6 + 0 flagellum of *Lecudina tuzetae*. *Cell Motil.* 4:369–383.

Goodenough, U. W. 1983. Motile detergent-extracted cells of *Tetrahymena* and *Chlamydomonas*. *J. Cell Biol.* 96:1610–1621.

Goodenough, U., and Heuser, J. 1985. Outer & inner dynein arms of cilia and flagella. *Cell* 41:341–342.

Graham, L. E., Graham, J. M., and Kranzfelder, J. A. 1986. Irradiance, daylength and temperature effects on zoosporogenesis in *Coleochaete scutata* (Charophyceae). *J. Phycol.* 22:35–39.

Gray, J. 1955. The movement of sea-urchin spermatazoa. *J. Exp. Biol.* 32:775–801.

Gray, J., and Hancock, G. J. 1955. The propulsion of sea-urchin spermatazoa. *J. Exp. Biol.* 32:802–814.

Green, J. C. 1973. Studies in the fine structure and taxonomy of flagellates in the genus *Pavlova*. II. A freshwater representative, *Pavlova granifera* (Mack) comb. nov. *Br. phycol. J.* 8:1–12.

Green, J. C. 1976. Notes on the flagellar apparatus and taxonomy of *Pavlova mesolychnon* van der Veer, and on the status of *Pavlova* Butcher and related genera within the Haptophyceae. *J. mar. biol. Assoc. U.K.* 56:595–602.

Green, J. C. 1980. The fine structure of *Pavlova pinguis* Green and a preliminary survey of the order Pavlovales (Prymnesiophyceae). *Br. phycol. J.* 15:151–191.

Green, J. C., and Course, P. A. 1983. Extracellular calcification in *Chrysotila lamellosa* (Prymnesiophyceae). *Br. phycol. J.* 18:367–382.

Green, J. C., and Hibberd, D. J. 1977. The ultrastructure and taxonomy of *Diacronema vlkianum* (Prymnesiophyceae) with special reference to the haptonema and flagellar apparatus. *J. mar. biol. Assoc. U.K.* 57:1125–1136.

Green, J. C., and Manton, I. 1970. Studies in the fine structure and taxonomy of flagellates in the genus *Pavlova*. I. A revision of *Pavlova gyrans*, the type species. *J. mar. biol. Assoc. U.K.* 50:1113–1130.

Greenwood, A. D. 1959. Observations on the structure of the zoospores of *Vaucheria*, II, *J. Exp. Bot.* 10:55–68.

Greenwood, A. D., Manton, I., and Clarke, B. 1957. Observations on the structure of the zoospores of *Vaucheria*. *J. Exp. Bot.* 8:71–86.

Hall, R. P. 1925. Binary fission in *Oxyrrhis marina* Dujardan. *Univ. Calif. Publ. Zool.* 26:281–324.

Hancock, G. J. 1953. Self-propulsion of microscopic organisms through liquids. *Proc. Roy. Soc. A* 217:96–121.

Hand, W. G., Collard, P. A., and Davenport, D. 1965. The effects of temperature and salinity changes on swimming rate in the dinoflagellates, *Gonyaulax* and *Gyrodinium*. *Biol. Bull.* 128:90–101.

Hand, W. G., and Haupt, W. 1971. Flagellar activity of the colony members of *Volvox aureus* Ehrbg. during light stimulation. *J. Protozool.* 18:361–364.

Hand, W. G., and Schmidt, J. A. 1975. Phototactic orientation by the marine dinoflagellate *Gyrodinium dorsum* Kofoid. II. Flagellar activity and overall response mechanism. *J. Protozool.* 22:494–498.

Hara, Y., and Chihara, M. 1985. Ultrastructure and taxonomy of *Fibrocapsa japonica* (Class Raphidophyceae). *Arch. Protistenk.* 130:133–141.

Hasle, G. R. 1950. Phototactic vertical migration in marine dinoflagellates. *Oikos* 2:162–176.

Hasle, G. R. 1964. More on phototactic vertical migration in marine dinoflagellates. *Nytt Mag. Bot.* 2:139–147.

Heath, J. B., and Darley, W. M. 1972. Observations on the ultrastructure of the male gamete of *Biddulphia levis* Ehr. *J. Phycol.* 8:51–59.

Henry, E. C. 1987. Primitive reproductive characters and a photoperiodic response in *Saccorhiza dermatodea* (Laminariales, Phaeophyceae). *Br. phycol. J.* 22:23–31.

Henry, E. C., and Cole, K. M. 1982. Ultrastructure of swarmers in the Laminariales (Phaeophyceae). *J. Phycol.* 18:550–569.

Herman, E. M., and Sweeney, B. M. 1977. Scanning electron microscopic observations of the flagellar structure of *Gymnodinium splendens* (Pyrrophyta, Dinophyceae). *Phycologia* 16:115–118.

Herth, W. 1982. Twist (and rotation?) of central-pair microtubules in flagella of *Poterioochromonas*. *Protoplasma* 112:17–25.

Heywood, P. 1972. Structure and origin of flagellar hairs in *Vacuolaria virescens*. *J. Utrastruct. Res.* 39:608–623.

Hibberd, D. J. 1973. Observations on the ultrastructure of flagellar scales in the genus *Synura* (Chrysophyceae). *Arch. Mikrobiol.* 89:291–304.

Hibberd, D. J., and Leedale, G. F. 1970. Eustigmatophyceae—a new algal class with unique organization of the motile cell. *Nature* 225:758–760.

Hilenski, L. L., and Walne, P. L. 1985. Ultrastructure of the flagella of the colorless phagotroph *Peranema trichophorum* (Euglenophyceae). I. Flagellar mastigonemes. *J. Phycol.* 21:114–125.

Hill, D. R. A., and Wetherbee, R. 1986. *Proteomonas sulcata* gen. et sp. nov. (Cryptophyceae), a cryptomonad with two morphologically distinct and alternating forms. *Phycologia* 25:521–543.

Hines, M., and Blum, J. J. 1978. Bend propagation in flagella. I. Derivation of equations of motion and their simulation. *Biophys. J.* 23:41–57.

Hines, M., and Blum, J. J. 1979. Bend propagation in flagella. II. Incorporation of dynein cross-bridge kinetics into the equations of motion. *Biophys. J.* 25:421–441.

Hoffman, L. R. 1960. Chemotaxis of Oedogonium sperms. *Southwest. Nat.* 5:111–116.

Hoffman, L. R. 1973. Fertilization in Oedogonium. I. Plasmogamy. *J. Phycol.* 9:62–84.

Hoffman, L. R., and Manton, I. 1962. Observations on the fine structure of the zoospore of Oedogonium cardiacum with special reference to the flagellar apparatus. *J. Exp. Bot.* 13:443–449.

Hoffman, L. R., and Manton, I. 1963. Observations on the fine structure of the zoospore of Oedogonium. II. The spermatozoid of O. cardiacum. *Am. J. Bot.* 50:455–463.

Hogg, J. 1855. Cilia in Diatomaceae. *J. Microsc. Sci.* 3:235–236.

Höhfeld, I., Otten, J., and Melkonian, M. 1988. Contractile eukaryotic flagella: Centrin is involved. *Protoplasma* 147:16–24.

Holwill, M. E. J. 1965. The motion of Strigomonas oncopelti. *J. Exp. Biol.* 42:125–137.

Holwill, M. E. J. 1966a. The motion of Euglena viridis: The role of flagella. *J. Exp Biol.* 44:579–588.

Holwill, M. E. J. 1966b. Physical aspects of flagellar movement. *Physiol. Rev.* 46:696–785.

Holwill, M. E. J. 1982. Dynamics of eukaryotic flagellar movement. In W. B. Amos and J. G. Duckett (eds.), *Prokaryotic and Eukaryotic Flagella*. Cambridge University Press, New York, pp. 289–312.

Holwill, M. E. J., and Peters, P. D. 1974. Dynamics of the hispid flagellum of Ochromonas danica. *J. Cell Biol.* 62:322–328.

Holwill, M. E. J., and Sleigh, M. A. 1967. Propulsion of hispid flagella. *J. Exp. Biol.* 47:267–276.

Hoops, H. J., and Witman, G. B. 1983a. Flagellar motion in the green alga Chlorogonium elongatum in the presence and absence of basal bodies and associated structures. *J. Cell Biol. (Suppl.)* 97:194a.

Hoops, H. J., and Witman, G. B. 1983b. Outer doublet heterogeneity reveals structural polarity related to beat direction in Chlamydomonas flagella. *J. Cell Biol.* 97:902–908.

Hoops, H. J., Wright, R. L., Jarvik, J. W., and Witman, G. B. 1984. Flagellar waveform and rotational orientation in a Chlamydomonas mutant lacking normal striated fibers. *J. Cell Biol.* 98:818–824.

Hori, T., and Moestrup, Ø. 1987. Ultrastructure of the flagellar apparatus in Pyramimonas octopus (Prasinophyceae). I. Axoneme structure and numbering of peripheral doublets/triplets. *Protoplasma* 138:137–148.

Hosokawa, Y., and Miki-Noumura, T. 1987. Bending motion of Chlamydomonas axonemes after extrusion of central-pair microtubules. *J. Cell Biol.* 105:1297–1301.

Hyams, J. S. 1982. The Euglena paraflagellar rod: Structure, relationship to other flagellar components and preliminary biochemical characterization. *J. Cell Sci.* 55:199–210.

Hyams, J. S., & Borisy, G. G. 1975. The dependence of the waveform and direction of beat of Chlamydomonas flagella on calcium ions. *J. Cell Biol. (Suppl.)* 67:186a.

Hyams, J. S., and Borisy, G. G. 1978. Isolated flagellar apparatus of *Chlamydomonas:* Characterization of forward swimming and alteration of waveform and reversal of motion by calcium ions *in vitro. J. Cell Sci.* 33:235–253.

Jahn, T. L., and Bovee, E. C. 1964. Protoplasmic movements and locomotion of Protozoa. In S. H. Hunter (ed.), *Physiology and Biochemistry of Protozoa,* Vol. III. Academic Press, New York, pp. 61–129.

Jahn, T. L., and Bovee, E. C. 1967. Motile behavior of protozoa. In T.-T. Chen (ed), *Research in Protozoology,* Vol. I. Pergamon Press, New York, pp. 41–200.

Jahn, T. L., and Bovee, E. C. 1968. Locomotive and motile response in *Euglena.* In D. E. Buetow (ed.), *The Biology of Euglena,* Vol. I. Academic Press, New York, pp. 45–108.

Jahn, T. L., Fonseca, J. R., and Landman, M. 1963a. Mechanisms of locomotion of flagellates. III. *Peranema, Petalomonas* and *Entosiphon. J. Protozool.* (Suppl.) 10:11.

Jahn, T. L., Harmon, W. M., and Landman, M. 1963b. Mechanisms of locomotion in flagellates. I. *Ceratium. J. Protozool.* 10:358–363.

Jahn, T. L., Bovee, E. C., Fonseca, J. R., and Landman, M. 1964a. Mechanism of flagellate locomotion. *Proc. 10th Int. Bot. Cong.,* p. 508.

Jahn, T. L., Landman, M. D., and Fonseca, J. R. 1964b. The mechanism of locomotion of flagellates. II. Function of the mastigonemes of *Ochromonas. J. Protozool.* 11:291–296.

Jarosch, R. 1970. Über die Geisselwellen von *Synura bioretti* und die Mechanik des uniplanaren Wellenschlags. *Protoplasma* 69:201–214.

Jarosch, R. 1972. The participation of rotating fibrils in biological movements. *Acta Protozool.* 11:23–28.

Jarosch, R., and Fuchs, B. 1975. Zur Fibrillenrotation in der *Synura*-Geissel. *Protoplasma* 85:285–290.

Johnston, D. N., Silvester, N. R., and Holwill, M. E. J. 1980. An analysis of the shape and propagation of waves on the flagellum of *Crithidia oncopelti. J. Exp. Biol.* 80:299–315.

Kalinsky, R. G. 1971. *Pedinomonas minor* Korschikoff, a rare alga in Ohio. *J. Phycol.* 7:82–83.

Kamiya, R., and Hasegawa, E. 1987. Intrinsic difference in beat frequency between the two flagella of *Chlamydomonas reinhardtii. Exp. Cell Res.* 173:299–304.

Kamiya, R., and Okamoto, M. 1985. A mutant of *Chlamydomonas reinhardtii* that lacks the flagellar outer dynein arms but can swim. *J. Cell Sci.* 74:181–191.

Kamiya, R., and Witman, G. B. 1984. Submicromolar levels of calcium control the balance of beating between the two flagella in demembranated models of *Chlamydomonas. J. Cell Biol.* 98:97–107.

King, J. M. 1973. *Gloeococcus minutissimus* sp. nov. isolated from soil. *J. Phycol.* 9:349–352.

Klyver, F. D. 1929. Notes on the life history of *Tetraspora gelatinosa* (Vauch.) Desv. *Arch. Protistenk.* 66:290–296.

Kofoid, C. A. 1906. On the significance of the asymmetry in *Tripsolenia. Univ. Calif. Publ. Zool.* 3:127–133.

Kofoid, C. A., and Swezy, O. 1921. The free-living unarmored Dinoflagellata. *Mem. Univ. Calif.* 5:vii, 1–562.

Kugrens, P., and Lee, R. E. 1988. Ultrastructure of fertilization in a cryptomonad. *J. Phycol.* 24:385–393.

Kugrens, P., Lee, R. E., and Andersen, R. A. 1987. Ultrastructural variations in cryptomonad flagella. *J. Phycol.* 23:511–518.

Leadbeater, B., and Dodge, J. D. 1967a. An electron microscope study of dinoflagellate flagella. *J. Gen. Microbiol.* 46:305–314.

Leadbeater, B., and Dodge, J. D. 1967b. Fine structure of the dinoflagellate transverse flagellum. *Nature* 213:421–422.

Lee, K. W., and Bold, H. C. 1973. *Pseudocharaciopsis texensis* gen. et sp. nov., a new member of the Eustigmatophyceae. *Br. phycol. J.* 8:31–37.

Lee, R. E., and Kugrens, P. 1986. The occurrence and structure of flagellar scales in some freshwater cryptophytes. *J. Phycol.* 22:549–552.

Leedale, G. F. 1967. *Euglenoid Flagellates.* Prentice-Hall, Englewood Cliffs, N.J., xiii + 242 pp.

Leedale, G. F., Leadbeater, B. S. C., and Massalski, A. 1970. The intracellular origin of flagellar hairs in the Chrysophyceae and Xanthophyceae. *J. Cell Sci.* 6:701–719.

Leedale, G. F., Meeuse, B. J. D., Pringsheim, E. G. 1965. Structure and physiology of *Euglena spirogyra. Arch. Mikrobiol.* 50:68–102.

Lembi, C. A., and Herndon, W. R. 1966. Fine structure of the pseudocilia of *Tetraspora. Can J. Bot.* 44:710–712.

Loiseaux, S. 1973. Ultrastructure of zoidogenesis in unilocular zoidocysts of several brown algae. *J. Phycol.* 9:277–289.

Lowndes, A. G. 1936. Flagella movement. *Nature* 138:210–211.

Lowndes, A. G. 1941. On flagellar movement in various organisms. *Proc. Zool. Soc. Lond.* A 111:111–134.

Lowndes, A. G. 1943. The swimming of unicellular flagellate organisms. *Proc. Zool. Soc. Lond.* A 113:99–107.

Lowndes, A. G. 1944. The swimming of *Monas stigmatica* and *Peranema trichophorum* (Ehrbg.) Stein and *Volvox* sp. Additional experiments on the working of a flagellum. *Proc. Zool. Soc. Lond.* A 114:325–338.

Luck, D. J. L., Huang, B., and C. J. Brokaw, 1982. A regulatory mechanism for flagellar function is revealed by suppressor analysis in *Chlamydomonas. Cell Motil. (Suppl.)* 1:159–164.

McFadden, G. I., Schulze, D., Surek, B., Salisbury, J. L., and Melkonian, M. 1987. Basal body reorientation mediated by a Ca^{2+}-modulated contractile protein. *J. Cell Biol.* 105:903–912.

Maier, I. 1984. *Sexualität bei Braunalgen aus der Ordnung Laminariales.* Hartung-Gorre Verlag, Konstanz, 126 pp.

Maier, I., and Müller, D. G. 1986. Sexual pheromones in algae. *Biol. Bull.* 170:145–175.

Manton, I. 1952. The fine structure of plant cilia. In *Structural Aspects of Cell Physiology (Symp. Soc. Exp. Biol. VI).* Cambridge University Press, London, pp. 306–319.

Manton, I. 1959a. Observations on the internal structure of the spermatozoid of *Dictyota. J. Exp. Bot.* 10:448–461.

Manton, I. 1959b. Electron microscopical observations on a very small flagellate: The problem of *Chromulina pusilla* Butcher. *J. mar. biol. Assoc. U.K.* 38:319–333.

Manton, I., and Clarke B. 1951. Electron microscope observations on the zoospores of *Pylaiella* and *Laminaria. J. Exp. Bot.* 2:242–246.

Manton, I., and Clarke, B. 1956. Observations with the electron microscope on the internal structure of the spermatozoid of *Fucus. J. Exp. Bot.* 7:416–432.

Manton, I., Clarke B., and Greenwood, A. D. 1953. Further observations with the electron microscope on spermatozoids in the brown algae. *J. Exp. Bot.* 4:319–329.

Manton, I., and Parke, M. 1960. Further observations on small green flagellates with special reference to possible relatives of *Chromulina pusilla* Butcher. J. mar. biol. Assoc. U.K. 39:275–298.

Manton, I., and Parke, M. 1962. Preliminary observations on scales and their mode of origin in *Chrysochromulina polylepis* sp. nov. *J. mar. biol. Assoc. U.K.* 42:565–578.

Manton, I., and von Stosch, H. A. 1966. Observations on the fine structure of the male gamete of the marine centric diatom *Lithodesmium undulatum. J. Roy. Microsc. Soc.* 85:119–134.

Marchant, H. J., Pickett-Heaps, J. D., and Jacobs, K. 1973. An ultrastructural study of the sporogenesis and the mature zoospore of *Klebsormidium flaccidum. Cytobios* 8:95–107.

Marchese-Ragona, S. P., Glazzard, N., and Holwill, M. E. J. 1983. A comparison of the wave parameters of flagellar axonemes with 9 + 2 and 9 + 1 microtubular configurations. *J. Submicrosc. Cytol.* 15:55–59.

Markey, D. R., and Bouck, G. B. 1977. Mastigoneme attachment in *Ochromonas. J. Ultrastruct. Res.* 59:173–177.

Maruyama, T. 1981. Motion of the longitudinal flagellum in *Ceratium tripos* (Dinoflagellida): A retractile flagellar motion. *J. Protozool.* 28:328–336.

Maruyama, T. 1985a. Extraction model of the longitudinal flagellum of *Ceratium tripos* (Dinoflagellida): Reactivation of flagellar retraction. *J. Cell Sci.* 58:109–123.

Maruyama, T. 1985b. Ionic control of the longitudinal flagellum in *Ceratium tripos* (Dinoflagellida). *J. Protozool.* 32:106–110.

Mast, S. O. 1926. Reactions to light in *Volvox*, with special reference to the process of orientation. *Z. Vergl. Physiol.* 4:637–658.

Melkonian, M. 1979. An ultrastructural study of the flagellate *Tetraselmis cordiformis* Stein (Chlorophyceae) with emphasis on the flagellar apparatus. *Protoplasma* 98:139–151.

Melkonian, M. 1980. Flagellar roots, mating structure and gametic fusion in the green alga *Ulva lactuca* (Ulvales). *J. Cell. Sci.* 46:149–169.

Melkonian, M. 1981. The flagellar apparatus of the scaly green flagellate *Pyramimonas obovata:* Absolute configuration. *Protoplasma* 108:341–355.

Melkonian, M. 1982. Effect of divalent cations on flagellar scales in the green flagellate *Tetraselmis cordiformis. Protoplasma* 111:221–233.

Melkonian, M., and Preisig, H. R. 1982. Twist of central pair microtubules in the flagellum of the green flagellate *Scourfieldia caeca. Cell Biol. Int. Rep.* 6:269–277.

Melkonian, M., and Preisig, H. R. 1984. Ultrastructure of the flagellar apparatus in the green flagellate *Spermatozopsis similis. Plant Syst. Evol.* 146:145–162.

Melkonian, M., and Preisig, H. R. 1986. A light and electron microscopic study of *Scherffelia dubia*, a new member of the scaly green flagellates (Prasinophyceae). *Nord. J. Bot.* 6:235–256.

Melkonian, M., Reize, I. B., and Preisig, H. R. 1987. Maturation of a flagellum/basal body requires more than one cell cycle in algal flagellates: Studies on *Nephroselmis olivacea* (Prasinophyceae). In W. Wiessner, D. G. Robinson, and R. C. Starr (eds.), *Algal Development*. Springer-Verlag, Berlin, pp. 102–113.

Melkonian, M., Robenek, H., and Rassat, J. 1982. Flagellar membrane specializations and their relationship to mastigonemes and microtubules in *Euglena gracilis. J. Cell Sci.* 55:115–135.

Metzner, P. 1929. Bewegungsstudien an Peridineen. *Z. Bot. (Z. Pflantz. Physiol.)* 22:225–265.

Mignot, J.-P. 1967. Structure et ultrastructure de quelques Chloromonadines. *Protistologica* 3:5–23.

Mignot, J.-P. 1976. Compléments à l'étude des chloromonadines: Ultrastructure de *Chattonella subsalsa* Biecheler flagellé d'eau saumâtre. *Protistologica* 12:279–293.

Miles, C. A., and Holwill, M. E. J. 1969. Asymmetric flagellar movement in relation to the orientation of the spore of *Blastocladiella emersonii*. *J. Exp. Biol.* 50:683–687.

Miller, R. L. 1966. Chemotaxis during fertilization in the hydroid *Companularia*. *J. Exp. Zool.* 162:23–44.

Miller, R. L. 1970. Sperm migration prior to fertilization in the hydroid *Gonothyrea loveni*. *J. Exp. Zool.* 175:493–504.

Miller, R. L. 1975. Chemotaxis of the spermatozoa of *Ciona intestinalis*. *Nature* 254;244–245.

Miller, R. L. 1976. Flagellar wave morphology during animal sperm chemotaxis. *J. Cell. Biol., Abstr. Int. Congr. Cell Biol.* 70:341a.

Miller, R. L. 1977. Chemotactic behavior of the sperm of chitons (*Mollusca, Polyplacophora*). *J. Exp. Biol.* 202:203–212.

Miller, R. L. 1979. Sperm chemotaxis in the hydromedusae. I. Species-specificity and sperm behaviour. *Mar. Biol.* 53:99–114.

Miller, R. L., and Brokaw, C. J. 1970. Chemotactic turning behaviour of *Tubularia* spermatozoa. *J. Exp. Biol.* 52:699–706.

Mitchell, D. R., and Rosenbaum, J. L. 1985. A motile *Chlamydomonas* flagella mutant that lacks outer dynein arms. *J. Cell Biol.* 100:1228–1234.

Moestrup, Ø. 1970a. On the fine structure of the spermatozooids of *Vaucheria sescuplicaria* and on the later stages of spermatogenesis. *J. mar. biol. Assoc. U.K.* 50:513–523.

Moestrup, Ø. 1970b. The fine structure of spermatozoids of *Chara corallina*, with special reference to microtubules and scales. *Planta* 93:295–308.

Moestrup, Ø. 1972. Observations on the fine structure of spermatozoids and vegetative cells of the green algae *Golenkinia*. *Br. phycol J.* 7:169–183.

Moestrup, Ø. 1982. Flagellar structure in algae: a review, with new observations particularly on the Chrysophyceae, Pheophyceae (Fucophyceae), Euglenophyceae, and *Reckertia*. *Phycologia* 21:427–528.

Moestrup, Ø. 1983. Further studies on *Nephroselmis* and its allies (Prasinophyceae). I. The question of the genus *Bipedinomonas*. *Nord. J. Bot.* 3:609–627.

Moestrup, Ø., and Ettl, H. 1979. A light and electron microscopical study of *Nephroselmis olivacea* (Prasinophyceae). *Opera Bot.* 49:1–39.

Moestrup, Ø., and Hoffman, L. R. 1975. A study of the spermatozoids of *Dichotomosiphon tuberosus* (Chlorophyceae). *J. Phycol.* 11:225–235.

Moestrup, Ø., Hori, T., and Kristiansen, A. 1987. Fine structure of *Pyramimonas octopus* sp. nov., an octoflagellated benthic species of *Pyramimonas* (Prasinophyceae), with some observations on its ecology. *Nord. J. Bot.* 7:339–352.

Moestrup, Ø., and Thomsen, H. A. 1986. Ultrastructure and reconstruction of the flagellar apparatus in *Chrysochromulina apheles* sp. nov. (Pymnesiophyceae = Haptophyceae). *Can. J. Bot.* 64:593–610.

Moewus, F. 1933. Untersuchungen über die Sexualität und Entwicklung von Chlorophyceen. *Arch. Protistenk.* 80:469–526.

Moore, A. R. 1916. The mechanism of orientation in *Gonium*. *J. Exp. Zool.* 21:431–432.

Moore, A. R., and Goodspeed, T. H. 1911. Galvanotropic orientation in *Gonium pectorale*. *U. Calif. Publ. Physiol.* 4:17–23.

Morel-Laurens, N. 1987. Calcium control of phototactic orientation in *Chlamydomonas reinhardtii:* Sign and strength of response. *Photochem. Photobiol.* 45:119–128.

Mosto, P. 1978. *Heterococcus tellii* sp. nov. (Xanthophyceae). *Physis Secc. B Aguas Cont. Org.* 38:31–35.

Müller, D. G. 1978. Locomotive responses of male gametes to the species specific sex attractant in *Ectocarpus siliculosus* (Pheaophyta). *Arc. Protistenk.* 120:371–377.

Müller, D. G., and Falk, H. 1973. Flagellar structure of the gametes of *Ectocarpus siliculosus* (Phaeophyta) as revealed by negative staining. *Arch. Mikrobiol.* 91:313–322.

Müller, D. G., and Lüthe, N. M. 1981. Hormonal interaction in sexual reproduction of *Desmarestia aculeata (Phaeophyceae). Br. phycol. J.* 16:351–356.

Müller, D. G., Clayton, M. N., and Germann, I. 1985a. Sexual reproduction and life history of *Perithalia caudata* (Sporochnales, Phaeophyta). *Phycologia* 24:467–473.

Müller, D. G., Maier, I., and Gassmann, G. 1985b. Survey on sexual pheromone specificity in Laminariales (Phaeophyceae). *Phycologia* 24:475–484.

Nichols, K. M. and Rikmenspoel, R. 1977. Mg^{2+}-dependent electrical control of flagellar activity in *Euglena. J. Cell Sci.* 23, 211–225.

Nichols, K. M., and Rikmenspoel, R. 1978a. Control of flagellar motion in *Chlamydomonas* and *Euglena* by mechanical microinjection of Mg^{2+} and Ca^{2+} and by electric current injection. *J. Cell Sci.* 29:233–247.

Nichols, K. M., and Rikmenspoel, R. 1978b. Control of flagellar motility in *Euglena* and *Chlamydomonas*. Microinjection of EDTA, EGTA, Mn^{2+}, and Zn^{2+}. *Exp. Cell Res.* 116:333–340.

Nichols, K. M., and Rikmenspoel, R. 1980. Flagellar waveform reversal in *Euglena. Exp. Cell Res.* 129:377–381.

Nichols, K. M., Jacklet, A., and Rikmenspoel, R. 1980. Effects of Mg^{2+} and Ca^{2+} on photoinduced *Euglena* flagellar responses. *J. Cell Biol.* 84:355–363.

Norris, R. E. 1980. Prasinophytes. In E. R. Cox (ed.), *Phytoflagellates.* Elsevier/North Holland, New York, pp. 85–145.

Nultsch, W. 1983. The photocontrol of movement of *Chlamydomonas*. In D.C. Cosens and D. Vince-Prue (eds.), *The Biology of Photoreception. Cambridge Soc. Exp. Biol. Symp.* 36:521–539.

Nultsch, W., and Häder, D.-P. 1988. Photomovement in motile microorganisms—II. *Photochem. Photobiol.* 47:837–869.

Okagaki, T., and Kamiya, R. 1986. Microtubule sliding in mutant *Chlamydomonas* axonemes devoid of outer or inner dynein arms. *J. Cell Biol.* 103:1895–1902.

Omoto, C. K., and Brokaw, C. J. Bending patterns of *Chlamydomonas* flagella: II. Calcium effects on reactivated *Chlamydomonas* flagella. *Cell Motil.* 5:53–60.

Omoto, C. K., Kung, C. 1980. Rotation and twist of the central-pair microtubules in the cilia of *Paramecium. J. Cell Biol.* 87:33–46.

Omoto, C. K., and Witman, G. B. 1981. Functionally significant central-pair rotation in a primitive eukaryotic flagellum. *Nature* 290:708–710.

Parke, M., Manton, I., and Clarke, B. 1955. Studies on marine flagellates. II. Three new species of *Chrysochromulina. J. mar. biol. Assoc. U.K.* 34:579–609.

Parke, M., Manton, I., and Clarke, B. 1956. Studies on marine flagellates. III. Three further species of *Chrysochromulina. J. mar. biol. Assoc. U.K.* 35:387–414.

Parke, M., Manton, I., and Clarke, B. 1958. Studies on marine flagellates. IV. Morphology and microanatomy of a new species of *Chrysochromulina. J. mar. biol. Assoc. U.K.* 37:209–228.

Parke, M., Manton, I., and Clarke, B. 1959. Studies on marine flagellates. V. Morphology and microanatomy of *Chrysochromulina strobilus* sp. nov. *J. mar. biol. Assoc. U.K.* 38:169–188.

Pascher, A. 1925. Heterokonte. In *Die Süsswasserflora Deutschlands, Österreichs und der Schweiz* 11:1–118. Wien.

Pennick, N. 1981. Flagellar scales in *Hemiselmis brunnescens* Butcher and *H. virescens* Droop (Cryptophyceae). *Arch. Protistenk.* 124:267–270.

Peters, N. 1929. Über Orts-un Geisselbewegung bei marinen Dinoflagellaten. *Arch. Protistenk.* 67:291–321.

Piccini, E., Albergoni, V., and Coppellotti, O. 1975. ATPase activity in flagella from *Euglena gracilis*. Localization of the enzyme and effects of detergents. *J. Protozool.* 22:331–335.

Pickett-Heaps, J. D. 1968. Ultrastructure and differentiation in *Chara (fibrosa)* IV. Spermatogenesis. *Aust. J. Biol. Sci.* 21:655–690.

Pickett-Heaps, J. D., and Marchant, H. J. 1972. The phylogeny of the green algae: a new proposal. *Cytobios* 6:255–264.

Pitelka, D. R., and Schooley, C. N. 1955. Comparative morphology of some protistan flagella. *Univ. Calif. Pub. Zool.* 61:79–128.

Pommerville, J. 1978. Analysis of gamete and zygote motility in *Allomyces*. *Exp. Cell Res.* 113:161–172.

Prasad, B. N., and Srivastava, P. N. 1963. Observations on the morphology, cytology, and asexual reproduction of *Schizomeris leibleinii*. *Phycologia* 2:148–156.

Preisig, H. R., and Melkonian, M. 1984. A light and electron microscopical study of the green flagellate *Spermatozopsis similis* spec. nova. *Plant Syst. Evol.* 146:57–74.

Prensier, G., Vivier, E., Goldstein, S. F., and Schrével, J. 1980. Motile flagellum with a "3 + 0" ultrastructure. *Science* 207:1493–1494.

Prescott, G. W. 1978. *The Algae: A Review.* Houghton Mifflin, Boston, 436 pp.

Raven, J. A., and Richardson, K. 1984. Dinophyte flagella: a cost-benefit analysis. *New Phytol.* 98:259–276.

Racey, T. J., Hallett, R., and Nickel, B. 1981. A quasi-elastic light scattering and cinematographic investigation of motile *Chlamydomonas reinhardtii*. *Biophys. J.* 35:557–571.

Rees, A. J. J., and Leedale, G. F. 1980. The dinoflagellate transverse flagellum: Three-dimensional reconstructions from serial sections. *J. Phycol.* 16:73–80.

Rietema, H. 1969. A new type of life history in *Bryopsis* (Chlorophycea, Caulerpales). *Acta Bot. Neerl.* 18:615–619.

Rietema, H. 1971. Life-history studies in the genus *Bryopsis* (Chlorophyceae) IV. Life-histories in *Bryopsis hypnoides* Lamx. from different points along the European coasts. *Acta Bot. Neerl.* 20:291–298.

Ringo, D. L. 1967. Flagellar motion and fine structure of the flagellar apparatus in *Chlamydomonas*. *J. Cell Biol.* 33:543–571.

Rüffer, U., and Nultsch, W. 1985. High-speed cinematographic analysis of the movement of *Chlamydomonas*. *Cell Motil. Cytoskel.* 5:251–263.

Rüffer, U., and Nultsch, W. 1987. Comparison of the beating of cis and trans flagella of *Chlamydomonas*. *Cell Motil. Cytoskel.* 7:87–93.

Sager, R, and Granick, S. 1954. Nutritional control of sexuality in *Chlamydomonas reinhardi*. *J. Gen. Physiol.* 37:729–742.

Sakaguchi, H., and Iwasa, K. 1979. Two photophobic responses in *Volvox carteri*. *Plant Cell Physiol.* 20:909–916.

Sakaguchi, H., and Tawada, K. 1977. Temperature effect on the photo-accumulation and phobic response of *Volvox aureus*. *J. Protozool.* 24:284–288.

Sakakibara, H., and Kamiya, R. 1989. Functional recombination of outer dynein arms with outer arm-missing flagellar axonemes of a *Chlamydomonas* mutant. *J. Cell Sci.* 92:77–83.

Salisbury, J. L., and Floyd, G. L. 1978. Calcium-induced contraction of the rhizoplast of a quadriflagellate green alga. *Science* 202:975–977.

Salisbury, J. L., Swanson, J. A., Floyd, G. L., Hall, R., and Maihle, N. J. 1981. Ultrastructure of the flagellar apparatus of the green alga *Tetraselmis subcordiformis*. *Protoplasma* 107:1–11.

Sandgren, C. D., and Flanagin, J. 1986. Heterothallic sexuality and density dependent encystment in the Chrysophycean alga *Synura petersenii* Korsch. *J. Phycol.* 22:206–216.

Santore, U. J. 1983. Flagellar and body scales in the Cryptophyceae. *Br. phycol. J.* 18:239–248.

Satir, P. 1963. Studies on cilia. The fixation of the metachronal wave. *J. Cell Biol.* 18:345–365.

Satir, P. 1965. Studies on cilia. II. Examination of the distal region of the ciliary shaft and the role of the filaments in motility. *J. Cell Biol.* 26:805–834.

Schmidt, J. A., and Eckert, R. 1976. Calcium couples flagellar reversal to photostimulation in *Chlamydomonas reinhardtii*. *Nature* 262:713–715.

Schnepf, E., and Deichgräber, G. 1969. Über die Feinstruktur von *Synura petersenii* unter besonderer Berücksichtigung der Morphogenese ihrer Kieselschuppen. *Protoplasma* 68:85–106.

Schrével, J., and Besse, C. 1975. Un type flagellaire fonctionnel de base 6 + 0. *J. Cell Biol.* 66: 492–507.

Schultz, M. E., and Trainor, F. R. 1968. Production of male gametes and auxospores in the centric diatoms. *Cyclotella menaghiniana* and C. *cryptica*. *J. Phycol.* 4:85–88.

Schütt, F. 1895. Der Peridineen der Plankton-Expedition. I. Studien über die Zellen der Peridineen. *Ergebn D Plankton-Exped (Kiel und Leipzig)*. 4:1–170.

Segal, R. A., Huang, B., and Luck, D. J. L. 1983. Motility mutants of *Chlamydomonas* with altered wave forms. *J. Cell Biol.* 97:194a.

Segal, R. A., Huang, B., Ramanis, Z., and Luck, D. J. L. 1984. Mutant strains of *Chlamydomonas reinhardtii* that move backwards only. *J. Cell Biol.* 98: 2026–2034.

Segal, R. A., and Luck, D. J. 1985. Phosphorylation in isolated *Chlamydomonas* axonemes: A phosphoprotein may mediate the Ca^{2+}-dependent photophobic response. *J. Cell Biol.* 101:1702–1712.

Silvester, N. R., and Holwill, M. E. J. 1972. An analysis of hypothetical flagellar waveforms. *J. Theor. Biol.* 35:505–523.

Sleigh, M. A. 1964. Flagellar movement of the sessile flagellates *Actinomonas*, *Codonosiga*, *Monas* and *Poteriodendron*. *Quart. J. Microsc. Sci.* 105:405–414.

Sleigh, M. A. 1974. Patterns of movement of cilia and flagella. In M. A. Sleigh (ed.), *Cilia and Flagella*. Academic Press, London, pp. 79–92.

Sleigh, M. A. 1981. Flagellar beat patterns and their possible evolution. *Biosystems* 14:423–431.

Sluiman, H. J. 1983. The flagellar apparatus of the zoospore of the filamentous green alga *Coleochaete pulvinata*: absolute configuration and phylogenetic significance. *Protoplasma* 115:160–175.

Smyth, R. D., and Berg, H. C. 1982. Change in flagellar beat frequency of *Chlamydomonas* in response to light. *Cell Motil. (Suppl.)* 1:211–215.

Soyer, M.-O., Prévot, P., de Billy, F., Jalanti, T., Flach, F., and Gautier, A. 1982. *Prorocentrum micans* E., one of the most primitive dinoflagellates: I. The complex flagellar apparatus as seen in scanning and transmission microscopy. *Protistologica* 18:289–298.

Spencer, L. B. 1971. A study of *Vacuolaria virescens* Cienkowski. *J. Phycol.* 7:274–279.

Stewart, K. D., and Mattox, K. R. 1975. Comparative cytology, evolution and classification of the green algae with some consideration of the origin of other organisms with chlorophylls a and b. *Bot. Rev.* 41:104–135.

von Stosch, H. A. 1951. Entwicklungsgeschichtliche Untersuchungen an zentrischen Diatomeen I. Die Auxosporenbildung von *Melosira varians*. *Arch. Mikrobiol.* 16:101–35.

von Stosch, H. A. 1973. Observations on vegetative reproduction and sexual life cycles of two freshwater dinoflagellates, *Gymnodinium pseudopalustre* Schiller and *Woloszynskia apiculata* sp. nov. *Br. phycol. J.* 8:105–134.

Summers, K. E., and Gibbons, I. R. 1971. Adenosine-triphosphate-induced sliding of tubules in trypsin-treated flagella of sea urchin sperm. *Proc. Natl. Acad. Sci. U.S.A.* 68:3092–3096.

Suzaki, T., and Williamson, R. E. 1986. Reactivation of euglenoid movement and flagellar beating in detergent-extracted cells of *Astasia longa*: Different mechanisms of force generation are involved. *J. Cell Sci.* 80:75–89.

Sweeney, B. M. 1971. Laboratory studies of a green *Noctiluca* from New Guinea. *J. Phycol.* 7:53–58.

Sweeney, B. M. 1976. *Pedinomonas noctilucae* (Prasinophyceae), the flagellate symbiotic in *Noctiluca* (Dinophyceae) in southeast Asia. *J. Phycol.* 12:460–464.

Tamm, S. L., and Horridge, G. A. 1970. The relation between the orientation of the central fibrils and the direction of beat in cilia of *Opalina*. *Proc. Roy. Soc. B* 175:219–233.

Taylor, F. J. R. 1975. Non-helical transverse flagella in dinoflagellates. *Phycologia* 14:45–47.

Taylor, F. J. R. 1980. On dinoflagellate evolution. Biosystems 13:65–108.

Taylor, G. I. 1952. Analysis of the swimming of long and narrow animals. *Proc. Roy. Soc. A* 214:158–183.

Throndsen, J. 1973. Motility in some marine nanoplankton flagellates. *Norw. J. Zool.* 21:193–200.

Toth, R. 1976. The release, settlement and germination of zoospores in *Chorda tomentosa*. *J. Phycol.* 12:222–233.

Turner, F. R. 1968. An ultrastructural study of plant spermatogenesis. Spermatogenesis in *Nitella*. *J. Cell Biol.* 37:370–393.

Ulehla, V. 1911. Ultramikroscopische Studien über Geisselbewegung. *Biol. Zbl.* 31:689–705.

van der Veer, J. 1969. *Pavlova mesolychnon* (Chrysophyta), a new species from the Tamar estuary, Cornwall. *Acta Bot. Neerl.* 18:496–510.

Verworn, M. 1909. *Allgemeine Physiologie*, 5th ed. Gustav Fischer, Jena, xvi + 742 pp.

Vora, M. R., Wolken, J. J., and Ahn, K. 1967. Biochemical studies of the flagella of *Euglena gracilis* Z. *J. Protozool.* (*Suppl*) 14:17.

Walker, P. J., and Walker, J. C. 1962. Movement of trypanosome flagella. *J. Protozool.* (*Suppl.*) 10:109.

Watson, M. W. 1975. Flagellar apparatus, eyespot and behavior of *Microthamnion kuetzingianum* (Chlorophyceae) zoospores, *J. Phycol.* 11:439–448.

Wawrik, F. 1960. Sexualität bei *Mallomonas fastiga* var. *kriegerii*. Arch. Protistenk. 104:542–544.

Wawrik, F. 1971. Zygoten und Cysten bei *Stenocalyx klasnetii* (Bourr.) Fott, *Stenocalyx inconstans* Schmid und *Chroomonas acuta* Utermöhl. *Nova Hedw.* 21:599–604.

Wenrich, D. H. 1924. Studies on *Euglenamorpha hegneri* N. G., N. SP., a euglenoid flagellate found in tadpoles. *Biol. Bull.* 47:149–175.

Wetherbee, R., Platt, S. J., Beech, P. L., and Pickett-Heaps, J. D. 1988. Flagellar transformation in the heterokont *Epipyxis pulchra* (Chrysophyceae): direct observations using image enhanced light microscopy. *Protoplasma* 145:47–54.

Wujek, D. E. 1968. Some observations on the fine structure of three genera in the Tetrasporaceae. *Ohio J. Sci.* 68:187–191.

Xiaoping, G., Dodge, J. D., and Lewis, J. 1989. Gamete mating and fusion in the marine dinoflagellate *Scrippsiella* sp. *Phycologia* 28:342–351.

Zimmermann, B. 1977. Flagellar and body scales in the chrysophyte *Mallomonas multiunca* Asmund. *Br. phycol. J.* 12:287–290.

Zingmark, R. G. 1970. Sexual reproduction in the dinoflagellate *Noctiluca miliaris* Suriray. *J. Phycol.* 6:122–126.

Chapter 5

Molecular Mechanism of Flagellar Movement

Ritsu Kamiya

Flagella and cilia are microtubule-based, whiplike organelles that cause water flow by their rapid undulating movements. These organelles are present in a variety of organisms ranging from protozoa to mammals, including many kinds of unicellular algae. One of the best-studied flagellar systems is that of *Chlamydomonas reinhardtii*, a chlorophyte. A variety of flagella-deficient mutants exist for this species. This chapter describes our present understanding of how flagella and cilia move, with an emphasis on the data from *Chlamydomonas*. In particular, I discuss how the axoneme (the internal structure of flagella and cilia) is organized, what properties the mechanochemical transducer (dynein) has, and how axonemal movement is regulated.

The Axoneme

The 9 + 2 Structure

The eukaryotic flagellum (or cilium) is a membrane-covered organelle with a diameter of about 0.2 μm and a length of the order of 10 μm. The mechanism that produces flagellar or ciliary beating is entirely contained within the internal structure, the axoneme; this can be clearly demonstrated by the fact that flagella or cilia demembranated with detergent, such as Triton X–100, beat almost normally if ATP is supplied (Hoffmann-Berling, 1955; Gibbons and Gibbons, 1972).

The flagellar membrane functions to sense and transmit signals from

155

the external environment and thereby regulate movement. In addition, in *Chlamydomonas* it displays a special type of motility called gliding—a phenomenon that enables flagellate cells to creep on solid substrata (Lewin, 1952; Bloodgood, 1989). Although these functions are important for the cell's tactic behavior and mating responses, here I focus on the function of the axoneme.

Axonemes of most flagella and cilia have a common structural design of two microtubules (the central pair) surrounded by nine microtubules. The nine tubules are called outer doublets because they are composed of two subtubules: one subtubule with a circular cross section (subtubule A) and one with a C-shaped cross section (subtubule B) (Fig. 5.1). This structure is called 9 + 2. It is not known why the somewhat peculiar 9 + 2 structure is conserved and widespread among many species of organisms, but it is likely that the ninefold symmetry is preserved because the flagellum grows from the basal body (Ringo, 1967), a kind of centriole whose ninefold structure is preserved in many kinds of cell. However, why centrioles have a ninefold symmetry remains a mystery.

The 9 + 2 arrangement of the microtubules is maintained by several bridging structures. Neighboring outer doublets are connected by interdoublet links and by inner and outer dynein arms, which are multiprotein assemblies responsible for force production. The central pair is held at the center by radial spokes. All these bridges and arms are arranged at precise intervals along the outer-doublet microtubule: one interdoublet link, two or three radial spokes (the precise number varies with species), four outer dynein arms, and three inner dynein arms occur within a repeating unit distance of 96 nm. The axoneme, therefore, has a highly regular structure and is probably the most ordered of all cellular organelles.

All these bridges connect two outer-doublet microtubules in a way that allows their relative movement in certain directions. Movement is possible because these bridges make both tight and loose (or dynamic) contacts with microtubules at their opposite ends. For example, the radial spokes are firmly attached at one end to outer doublets and loosely interact with the central pair at the other end (spoke head); the central pair, therefore, has the freedom to rotate within the axoneme. The interdoublet links also appear to make loose bridges (Warner, 1983).

Having such a complex structure, the axoneme is naturally made up of many kinds of proteins besides tubulin. Electrophoresis analyses on *Chlamydomonas* mutants have demonstrated that the axoneme contains no less than 200 peptides, including 17 for radial spokes, about 20 for inner and outer dynein arms, and about 22 for the central-pair complex (Luck, 1984). How the synthesis and assembly of these proteins are regulated is a fascinating problem of cell biology (Lefebvre and Rosenbaum, 1986).

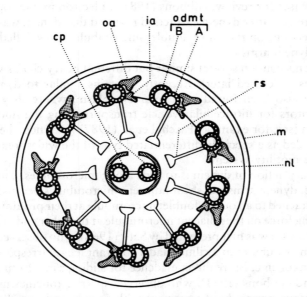

Figure 5.1. Schematic cross-section view of the 9 + 2 structure. cp, central-pair microtubule; oa, outer dynein arm; ia, iner dynein arm; odmt, outer-doublet microtubule [A, subtubule A; B, subtubule B]; rs, radial spoke; m, flagellar membrane; nl, nexin link (interdoublet link). Some minor variations have been observed in different species of axonemes. For example, in *Chlamydomonas* axonemes, one particular outer doublet lacks the outer dynein arm and three doublets contain beaklike structures within the subtubule B (Hoops and Witman, 1983). This structural heterogeneity among outer doublets may confer the bending direction on the axoneme.

Sliding Movement in Dynein–Tubulin System

It is now established that the driving force for axonemal beating is generated by interaction between dynein arms and microtubules, coupled with ATP hydrolysis. When dynein was first discovered by Gibbons in 1963 as an ATPase enzyme in *Tetrahymena* cilia, it was immediately regarded as the force producer for motility because ciliary motility had been known to require ATP [see for review, Gibbons (1981)]. Electron microscopy of axonemes before and after dynein extraction revealed that dynein was localized at two projections on the outer doublet microtubules, now called the inner and outer dynein arms.

Dynein arms contain several similar but different heavy chains with ATPase activities (discussed later). Recently, proteins similar to dynein heavy chains have also been found in cellular cytoplasm, where they probably work as motors for intracellular vesicle transport along the microtubular cytoskeleton [see, for example, Paschal et al. (1987b)]. Hence, the axoneme can be regarded as a specially differentiated type of the widespread dynein–microtubule motility systems.

The primary action of dynein is to slide along microtubules. Thus, within an axoneme, dynein arms can shear adjacent microtubules because the arms are stably attached to an outer-doublet microtubule at their proximal end and exert a sliding force on the adjacent microtubule at the other end. This sliding tubule mechanism was first presented by Satir (1968), who observed that outer-doublet microtubules in a cilium had a constant length irrespective of the bend in the axoneme. More direct evidence for sliding came from studies by Summers and Gibbons (1971), who observed sperm axonemes using dark-field microscopy after a brief treatment with trypsin and found that outer doublets actually slid out of the axoneme upon addition of ATP. They suggested that the long-distance sliding occurred because the trypsin treatment destroyed some axonemal structures such as interdoublet links that would normally restrict the microtubule sliding to a short distance. The outer doublets often slid against one another so extensively that the total length became as much as eight to nine times the original length. Hence, without doubt, adjacent outer doublets can slide apart in the presence of ATP.

Electron microscopy (Sale and Satir, 1977) further demonstrated that microtubule sliding occurs in a unidirectional manner; dynein arms on an outer doublet always "push" the adjacent one tipward, or, in other words, dynein arms themselves move toward the base of the outer doublet with which they interact. This polarity of movement is determined by an intrinsic property of dynein and the structural polarity of microtubules at the molecular level. The distal end of an outer doublet corresponds to the plus end of the cytoplasmic microtubule, which is defined as the faster-growing end. All kinds of dynein so far studied (including cytoplasmic types) have been shown to move toward the minus end of microtubules, whereas kinesin, another cytoplasmic microtubule motor, has been shown to move in the opposite direction (Vale et al., 1985).

Two important questions then arise as to how microtubule sliding is organized into axonemal beating, and, more fundamentally, how the microtubules slide at all. Although these questions have proved difficult and no definite answers have yet been obtained, recent studies have yielded various important findings that are relevant to these problems.

Central Pair and Radial Spokes

Flagella Can Beat Without Central Pair and Radial Spokes We know that dynein arms and microtubules are responsible for the primary force production process, but what do the central pair and radial spokes do in the axonemal beating mechanism? *Chlamydomonas* mutants lacking the central pair of radial spokes have been isolated and found to be nonmotile. It has also been shown that the outer doublets of these mutants can undergo sliding disintegration if axonemes are treated with trypsin and exposed to ATP (Witman et al., 1978). These observations seemed to indicate that the central-pair/radial-spoke structures are necessary for conversion of microtubule sliding into the bending movement of the axoneme. However, this idea was refuted following the observation that flagella of some organisms do not have the central-pair/radial-spoke system and yet are fully motile; for example, flagella of eel sperm (Gibbons et al., 1985). In fact, the central-pair/radial-spoke system was eventually shown to be dispensable for flagellar beating in *Chlamydomonas* also; Huang et al. (1982) found an unusual group of mutations called suppressors, which, when combined with central-pair deficient or radial-spoke deficient mutations, can restore motility *without* repairing the structural deficiency. Some of the suppressors were found to be mutations in dynein. Although it is not yet understood how the suppressors act on the paralyzed axonemes, it is clear that axonemes lacking the central pair or radial spokes are capable of beating.

Mutants that lack the central-pair/radial-spoke system and possess the suppressor mutations show abnormal flagellar beating in that beat frequency is lower and waveform is more symmetric in comparison to those in normal flagella (Brokaw et al., 1982). Therefore, the central-pair/radial-spoke system may be important for axonemes to beat efficiently with proper waveforms. Since some of the suppressor mutations are mutations in dynein, the central-pair/radial-spoke system in a normal axoneme probably acts to produce efficient axonemal boating by somehow regulating dynein–tubulin interactions.

Central-pair rotation

From the early days of research, the central pair has been postulated to be involved in the determination of the direction of axonemal beat. Almost all the axonemes examined were found to bend in the direction perpendicular to the plane of the two central microtubules, which were believed to lie at a fixed orientation with respect to other axonemal structures. Later studies, however, have suggested that the orientation of the central pair is not fixed but rotates, or at least has the freedom to rotate within the axoneme. Although the physiological importance of central-pair rotation remains to be clarified, it is conceivable that the rotation facilitates the propagation of regular waves by the axoneme.

Omoto and Kung (1980), examining *Paramecium* cilia by electron microscopy, found that the orientation of the central pair varied continuously up to 360 degrees depending on the phase of the ciliary beating cycle. The obvious conclusion was that the central pair rotates 360 degrees per beat. This idea has been supported by the discovery of another example, the flagellum of a tiny marine alga, *Micromonas pusilla*. This alga has an unusually short shaft of 9 + 2 flagellum with a long central pair protruding from the axonemal rod, and it swims by rotating the helical central pair like a bacterium swimming by rotating flagella (Omoto and Witman, 1981). [It is assumed that the flagellum of this organism has a liquid membrane that can rotate with the central pair.] A rotating fibrous projection, most likely the central pair, has also been observed in *Synura* flagella (Jarosch and Fuchs, 1975).

Central-pair rotation was thought to be a property of axonemes that beat with a three-dimensional waveform. However, observations with detergent-extracted "cell models" suggest that the central pair can also rotate in *Chlamydomonas* axonemes whose beating is largely two-dimensional; the central pair often spontaneously comes out of the axoneme and continuously rotates around the axonemal axis in the presence of ATP (Fig. 5.2; Kamiya, 1982). The extruded central pair of this organism has a left-handed helical shape, as in *M. pusilla,* indicating that the central pair must be twisted and strained when contained within the axoneme. Twist in the central pair has been observed by electron microscopy in flagella of *Scourfieldia caeca* (Melkonian and Preisig, 1982) and *Poterioochromonas* (Herth, 1982). The tendency to assume a helical shape may be important for the function of the central pair, because a helical central pair could propagate a wave of curvature when rotating within the axoneme and thereby assist the outer doublets to propagate bending waves.

It has not been established whether central-pair rotation is a general feature of all eukaryotic flagella. In cilia of metazoa such as ctenophores, the

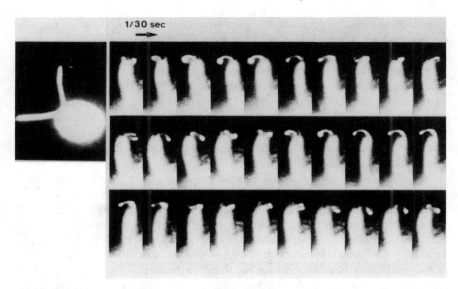

Figure 5.2. Rotation of the central-pair microtubules in a *Chlamydomonas* cell model. The central pair (the thin, curved protrusion seen at the axoneme tip) was spontaneously extruded from the axoneme and rotated at about 2 Hz in a reactivation medium containing a nonionic detergent and ATP. A series of video recordings taken at intervals of 1/30 s for 1 s (from upper left to lower right). The left photo shows a complete view of the cell carrying this axoneme. The flagellar length is about 11 μm. (From Kamiya, 1982, with permission.)

central pair has been reported not to rotate during the ciliary beat cycle (Tamm and Tamm, 1981); also, the central pair in sea urchin sperm flagella does not appear to rotate. A recent study by Gibbons et al. (1987), however, who held the head of a sperm with a micropipette and vibrated it while slowly rotating the vibration plane, indicated that the plane of flagellar beat can be rotated. Although the precise mechanism that determines the flagellar beat plane is to be established, it is possible that the central pair of sperm flagella can also rotate under certain conditions.

The mechanism of central-pair rotation is not yet understood. Nevertheless, a recent discovery that, under in vitro motility assay conditions, certain kinds of dynein can cause microtubules to rotate as well as to glide (Vale and Yano-Toyoshima, 1988) suggests that there may be a dyneinlike motor that rotates the central pair and may be located on the spoke head (discussed later).

Mechanism That Produces Flagellar Bending

As we have seen, axonemal beating takes place as a result of the relative sliding movement between outer doublets. This sliding should be precisely regulated in order to generate a regular waveform, because the waveform is determined by where, when, and how much the intertubule sliding takes place. Because dynein arms induce only unidirectional sliding of microtubules within the cylindrical structure, only dynein arms contained on one side of the axoneme (with respect to the bending plane running through the axonemal axis) should actively shear outer doublets; those contained on the other side should be at rest when the axoneme bends in one direction. The activation state should reverse when the axoneme bends in the opposite direction; the activated and resting states of each dynein arm should alternate when the axoneme beats.

What causes the oscillation? The most popular idea is that the axonemal mechanochemical system as a whole constitutes a feedback loop, with an assumption that the dynein–tubule interaction is regulated by a mechanical state of the axoneme (e.g., curvature), which is in turn determined by the extent of dynein–tubule interaction [for review, see Brokaw (1982)]. This kind of model seems to work reasonably well in computer simulations.

An oscillatory bending movement that is apparently effected through a feedback mechanism is observed in *Chlamydomonas* axonemes that have been spontaneously frayed in a reactivation solution containing ATP and detergent (Kamiya and Okagaki, 1986). The example in Figure 5.3 demonstrates that a pair of outer doublets bend and unbend at a frequency of 1–2 Hz in the presence of ATP, indicating that the essential mechanism for the oscillation is contained within outer doublets and dynein arms. Frame-by-frame analyses reveal that the mechanism for this movement can be understood by the assumption that the dynein–tubule interaction shears the two doublets unidirectionally, but that it is turned off at the site where curvature exceeds a certain critical value (for more details, see legend to Fig. 5.3). It is interesting to note that the critical curvature observed here is similar to the maximal curvature observed in the normal flagellar waveform; this suggests that the same turning-off mechanism is working in normal flagellar beating of a live cell.

Although the models based on mechanical feedback appear quite plausible, certain discrepancies have been found between computer models and real flagella. For example, real flagella do not change waveform greatly under high-viscosity conditions, whereas computer models easily become abnormal under such conditions (Brokaw, 1985). As Brokaw has suggested, beating mechanisms other than those controlled by curvature remain possible. Whatever the real mechanism is, however, it is most likely that the basis for the oscillation–wave propagation mechanism resides in the dynein–tubule interaction. Therefore, in order to understand the molecular mechanism of axonemal motility, it is essential to study the properties of dynein itself.

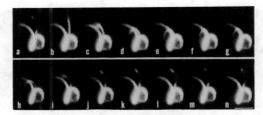

Figure 5.3. Bending movements displayed by two outer doublets in a frayed axoneme of *Chlamydomonas*. This axoneme has spontaneously detached from the cell body and split into three bundles of outer doublets in a reactivation solution containing ATP. The finest bundle of the three displayed a cyclical bending movement, whereas the other two remained stuck to the glass slide. Dark-field videotaped images taken every 1/15 s. The bending bundle split into two finer fibers, which are most likely single outer doublets. The cyclical bending movement can be interpreted by a mechanism that assumes the following: (1) the dynein arms on one microtubule pull the other tubule in a unidirectional manner; (2) the two tubules are firmly fixed at the base; thus, a bend is formed by the shear between the two tubules; (3) the dynein–tubule interaction is turned off where curvature exceeds a certain limit; and (4) microtubules are elastic, so they tend to assume a straight shape wherever dynein–tubule interaction is absent. (From Kamiya and Okagaki, 1986, with permission.)

Dynein

Structure

As mentioned earlier, *dynein* denotes a group of protein complexes that produce sliding force through interaction with microtubules. These complexes are composed of about 10 kinds of subunits, including two or three heavy chains with molecular masses in excess of 400 kD and enzyme activities that hydrolyze ATP.

Outer-Arm Dynein The best studied of all kinds of dynein is the axonemal outer-arm dynein, in particular, outer arms of sea urchin sperm, *Tetrahymena*, and *Chlamydomonas*. These dynein arms occur as projections located every 24 nm along the outer doublets, that is, at every third tubulin dimer. This kind of dynein can be extracted with high-salt-concentration solutions and purified by sucrose density-gradient centrifugation. The sedimentation coefficients are 21–23S.

When isolated and sprayed onto electron microscope grids, the outer arms show up as bouquetlike structures (Johnson and Wall, 1983; Goodenough and Heuser, 1984), that is, two- or three-headed structures bundled by a proximal stem (Fig. 5.4). The heads number two in sea urchin sperm and three in the other two species, each coinciding with the number of the high-molecular-mass subunits contained. It is not known how the bouquetlike outer arms are folded in situ in the axoneme, although most probably the proximal stems are attached to the A-subtubule of a doublet microtubule

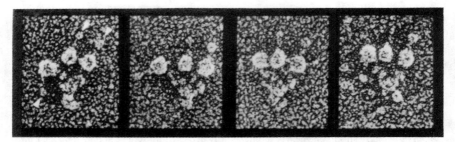

Figure 5.4. Isolated outer dynein arms observed by quick-freeze deep-etch electron microscopy. Arrowheads point to the fibrous projections from the three dynein heads. Magnification: 320,000. (From Goodenough and Heuser, 1989, with permission.)

and all the heads closely associated with the B-subtubule of the adjacent doublet.

An innovative electron microscope technique, called quick-freeze deep-etch microscopy, has revealed a thin, fibrous structure connecting each dynein head with the B-subtubule of the adjacent outer doublet (Goodenough and Heuser, 1984). This structure has been observed even under conditions where dynein is thought to be dissociated from the adjacent tubule; this observation is important with respect to the molecular mechanism of dynein function, although no theoretical models have considered it. Thin, fibrous projections, apparently the size of the connections observed in axonomes, have been observed in isolated outer dynein arms (see Fig. 5.4).

Chlamydomonas outer-arm dynein is composed of three high-molecular-mass peptides (α, β, and γ; MW = 415–480 kD), two intermediate chains with molecular masses of 69b kD and 78 kD, and 10 light chains of molecular masses ranging from 8 kD to 22 kD (Piperno and Luck, 1979b; Pfister et al., 1982; King and Witman, 1989). The identification of these peptides has been greatly facilitated by the isolation of mutants that lack the entire outer dynein arm (Huang et al., 1979). The composition of intermediate and light chains varies greatly among outer-arm dyneins from different species.

Inner-Arm Dynein The structure and biochemical properties of the inner-arm dynein are less clear than those of the outer arm. Explicitly or inexplicitly, the inner-arm has been assumed to be similar to the outer arm. However, recent studies indicate that the similarity cannot be taken too far. First of all, biochemical analyses of *Chlamydomonas* dynein-deficient mutants have shown that the protein compositions of the two kinds of arms differ greatly. Although the exact list of subunits that make up the inner arm is yet to be determined, the evidence so far obtained indicates that the inner arm contains five or more high-molecular-mass peptides, an intermediate chain with molecular mass 140 kD, and a few light chains. Interestingly, actin is contained in the inner-arm subunits (Piperno and Luck, 1979a). There is no common subunit in the two types of arms. The presence of five or more heavy chains indicates the presence of heterologous inner-arm species, because electron micrographs show that the size of the inner arm is not larger than that of the outer arm.

The inner and outer arm differ also in arrangement within the axoneme. Unlike outer arms, which are located at even intervals of 24 nm, inner arms are located at the proximal end of radial spokes, that is, at uneven intervals of 24, 32, and 40 nm. Moreover, there are two kinds of inner arms with different morphologies (Goodenough and Heuser, 1985). The presence of different species agrees with the previously mentioned biochemical data in favor of heterogeneity. More recent biochemical and electron microscopic analyses of mutant axonemes indicate that there may be three species of inner-arm dynein, each containing two heavy chains (Piperno et al., 1990). This means that a total of six inner-arm heavy chains may be present within an axoneme.

Despite the striking differences in organization and arrangement, the two kinds of arms share a common structural feature. Firstly, inner arms extracted from axonemes also look like bouquetlike particles in electron micrographs (Goodenough et al., 1987). Thus, the inner and outer arms have an overall structural similarity in molecular organization. Secondly, the intramolecular structures of all the heavy chains in both arms seem to be related. In 1986, Gibbons and his colleagues found that dynein heavy chains can be cleaved by irradiation with UV light in the presence of ATP and vanadate, a strong inhibitor of dynein ATPase. With the $A\alpha$ heavy chain of sea urchin outer-arm dynein, the cleavage yielded two peptides having almost identical molecular weights; the photochemical reaction seems to cleave the peptide at its mid-portion, where the ATP molecule presumably is bound (Lee-Eiford et al., 1986). Furthermore, it was found that the UV–vanadate treatment cleaves all kinds of heavy chains from outer arms, inner arms, and cytoplasmic dynein to yield peptides of similar size. Therefore, it is likely that these heavy chains share a common structural organization. More precise information on the primary structure of dynein subunits will soon become available, because cDNA for several heavy chains and intermediate chains has been cloned (Mitchell, 1989; Foltz and Asai, 1990).

Inner-Arm Versus Outer-Arm Functions

As we have seen, recent studies indicate that the inner and outer dynein arms differ greatly in structure and organization. Do they also differ, then, with respect to their functional roles?

A most important finding about the inner arm and outer-arm functions was obtained by Gibbons and Gibbons (1973). They showed that outer

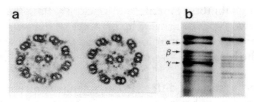

Figure 5.5. *Chlamydomonas* axonemes from wild type and *oda* mutant lacking the outer-arm dynein. (a) Cross-section electron micrographs of wild type and *oda*. (b) A portion of SDS polyacrylamide gel electrophoressi pattern showing the dynein heavy chains. α, β, and γ denote the three outer-arm heavy chains. A heavily stained band above the alpha is a band of a membrane protein. All six bands appearing in the *oda* pattern are thought to be inner-arm heavy chains.

dynein arms can be removed from demembranated sperm by extraction with high-salt-concentration solutions and that the extracted sperm axoneme beats in the presence of ATP at about half the frequency of that of the unextracted sperm. From these observations, Gibbons and Gibbons (1973) suggested that the two kinds of arms have almost identical functions, the effects being additive. Other work showed that the microtubule sliding rate in disintegrating axonemes is also halved upon removal of the outer arm (Yano and Miki-Noumura, 1981). These experiments demonstrated not only that the outer dynein arm is actually a mechanochemical motor for the axonemal beating, but also that axonemes can function without the outer arm. It is not clear, however, whether these results indicate any functional identity between the two kinds of arms, because it was not possible to remove the inner arm instead of the outer arm.

Studies using *Chlamydomonas* mutants, on the other hand, have suggested that the inner and outer arms differ somewhat in function. Mutants lacking the entire outer dynein arms and those lacking part of the inner arms have been isolated (Huang et al., 1979; Kamiya and Okamoto, 1985; Mitchell and Rosenbaum, 1985; Kamiya, 1988; Kamiya et al., 1989, 1991). The mutant lacking the outer arm (Fig. 5.5) swims at about one-half to one-third the velocity of the wild type, with an almost normal flagellar waveform but a reduced beat frequency (Brokaw and Kamiya, 1987). This example, therefore, also demonstrates that the outer dynein arm is not necessary for flagellar beating. However, the rate of microtubule sliding axonemes missing the outer arm was found to be as low as one-fifth to one-sixth that of the wild type (Okagaki and Kamiya, 1986). This indicates that the two types of arms differ, at least quantitatively, in their motility properties.

Evidence for a more striking difference between the inner and outer arms comes from a study on mutants deficient in inner arms. As stated before, the inner dynein arm of *Chlamydomonas* probably consists of three different subspecies (called I1, I2 and I3; Piperno et al., 1990). The inner-arm mutants now available are those lacking I1, I2, or both I1, and I2. It was found that mutants missing I1 and those missing I2 are both capable of swimming, although more slowly than the wild type. The reduced swimming rate is due to a reduced angle of flagellar bend rather than to any change in the beat frequency. Therefore, the lack of inner arms has a somewhat different effect on flagellar beating than the lack of outer arms (Brokaw and Kamiya, 1987). Interestingly, mutants missing both I1 and I2 are nonmotile; this suggests that flagella with a large defect in the inner arms cannot beat.

The preceding findings strongly argue against functional similarity between the two arms. Rather, the inner arms are apparently more important than the outer arms for generation of flagellar beat. Indeed, a considerable number of organisms have flagella that naturally lack the outer arms, whereas almost no examples are known in which inner arms are missing. For example, flagella of water fern sperm lack outer arms (Hyams, 1985). Why, then, do so many kinds of flagella have the outer arm at all? One possibility is that the presence of the outer arm not only enables a cell to swim faster, but also widens the repertoire of its behavior; for example, Chlamydomonas wild-type cells can swim both forward and backward, whereas outer-arm-lacking mutants have difficulty in swimming backward (Kamiya and Okamoto, 1985).

ATPase Cycle

The properties of dynein as an enzyme that hydrolyzes ATP have been investigated in only a small number of dynein species and, of these, Tetrahymena outer-arm dynein (22S dynein) is the most extensively studied (Johnson, 1985). These investigations have benefited greatly from research on myosin, the mechanochemical enzyme responsible for muscular contraction. ATP hydrolysis by dynein has been found to have a similar kinetic pathway to that of myosin, although the rate constants differ in corresponding kinetic steps. In the absence of ATP, actin and myosin form the so-called rigor complex and, similarly, dynein binds to microtubules. In both systems the firmly bound complex rapidly dissociates when ATP is added.

As in myosin, the kinetic pathway in dynein consists of rapid binding of ATP, fast ATP hydrolysis, and slow release of the products ADP and Pi. In order for dynein to work efficiently as a mechanochemical enzyme, ATP hydrolysis must proceed rapidly only when dynein works to cause microtubule sliding. Hence, the ATP hydrolysis cycle must be slow when dynein alone is present but greatly accelerated when dynein interacts with microtubules. This acceleration should be effected through an increase in rate of the slowest step, the step of product release. An acceleration of the ATPase cycle by microtubules has actually been observed in Tetrahymena outer-arm dynein by Omoto and Johnson (1986). These authors have shown that the

ATPase activity in 22S dynein (outer-arm dynein) increases 3–10 times when microtubules reconstructed from brain tubulin are added, up to a concentration of 50 mg/ml. The activation does not reach a plateau even with this high concentration of microtubules, suggesting that an extremely high concentration of microtubules is necessary for a full activation of dynein ATPase. It is not unreasonable that such a high concentration is needed, because the effective microtubule concentration in the vicinity of dynein arms in situ must be extremely high.

Kinetic properties of dynein other than that from *Tetrahymena* outer arms have been less extensively studied. Recent studies using in vitro motility assay technique (discussed later) indicate that there are significant differences in the properties among various dynein species. For example, with sea urchin dynein, brain microtubules attach to dynein-coated glass surfaces in the presence of ATP, rather than in its absence. Therefore, the kinetic pathway of ATP hydrolysis may differ in this and other species of dynein.

How ATP hydrolysis is coupled with force production in the dynein–microtubule system remains a matter of speculation. In the popular model for muscular contraction, myosin heads are postulated to undergo a cycle of conformational change accompanying association and dissociation with actin filaments. The cycle is assumed to be coupled with the ATP hydrolysis cycle. This cross-bridge model has been applied to explain the dynein–microtubule motility system and has served as a helpful working hypothesis. Recently, however, the cross-bridge mechanism of muscle contraction has been challenged. For example, there is evidence that actin moves along myosin as much as 100 nm upon hydrolysis of one ATP molecule—a distance much greater than the presumptive cross-bridge can span. (Yanagida et al., 1985). Another line of evidence suggests that myosin heads do not change their orientation during muscle contraction [see, for example, Cooke et al. (1982)]. Thus, the cross-bridge theory may be wrong at a rather fundamental level. It seems that we now require a new theory for the molecular mechanism of actin–myosin motility, as well as for dynein–microtubule motility.

Functional Assay of Dynein Subspecies

For a complete understanding of the flagellar motility mechanism, it is essential to clarify the specific function performed by each kind of dynein. This problem is being approached by analysis of mutants that lack specific dynein components [for review, see Kamiya et al. (1989)] An alternative approach, which does not depend on mutants, has recently been developed. This new method, the in vitro motility assay, examines various dynein preparations for the activity causing microtubule sliding by using a dynein-coated glass slide and microtubules reconstructed from mammalism tubulin (Paschal et al. 1987a; Vale and Yano-Toyoshima, 1988).

In the first example, sea urchin sperm outer-arm dynein (22S dynein) was shown, in the presence of ATP, to support gliding of microtubules at a maximal velocity of about 6 μm/s^{-1}, that is, at a velocity close to that of the microtubule sliding in disintegrating axonemes (Paschal et al., 1987a). This was the first demonstration that outer-arm dynein, on its own, can cause microtubule sliding. It was further shown that one (Aβ) of the two heavy chains of the 22S dynein is enough to cause this activity; actually, the sliding caused by the β-heavy chain was slightly faster than that caused by the entire outer arm (Sale and Fox, 1988). On the other hand, the other heavy chain, Aα, has not been observed to cause sliding.

Another study with *Tetrahymena* 14S dynein, a dynein species whose location in situ is not yet established, revealed a further striking phenomenon: microtubule rotation (Vale and Yano-Toyoshima, 1988). This kind of dynein was found to cause not only gliding of microtubules at a maximal speed of 2–3 μm^{-1}, but also a rotation of the microtubule around its axis. The rotation was linearly correlated with the translocation: about 1.8 rotations μm^{-1} on average. Although the physiological importance of the rotatory movement is not yet understood, this dynein may well have something to do with the rotation of the central pair. Alternatively, the rotatory force may play a role in the normal function of inner or outer arms.

Dynein preparations obtained from a *Chlamydomonas* mutant that lacks the outer arms have recently been found to have the ability to translocate microtubules in vitro (Kagami et al., 1990). At least three subfractions containing different sets of heavy chains caused translocation at 3–5 μm s^{-1}, that is, at a speed equal to the sliding rate in disintegrated axonemes of this mutant. Interestingly, one fraction was observed to induce microtubule sliding and rotation, similar to the movement observed with the 14S dynein of *Tetrahymena*. This raises the possibility that some inner-arm dynein can exert a torque on the outer doublets.

Microtubule gliding induced by dynein has been reported to occur only in a way such that microtubules proceed in the + direction, that is, in the direction corresponding to the base-to-tip direction of the axoneme. This direction is identical to that observed in sliding microtubules in disintegrating axonemes. It will be of great interest to see whether all the dynein subspecies within the axoneme always induce microtubule gliding in the same direction.

Regulation of Flagellar Movement

Many unicellular algae are attracted to bright environments but repelled by exceedingly strong light (Foster and Smyth, 1980). The strength of the photoresponse usually varies depending on the time of day and the growth conditions. Such complicated behavior, which must have evolved to optimize the living conditions of free-living photosynthetic organisms, is made possible by regulated flagellar movements. Previous studies have in fact indicated that the axonemal motility is generally regulated on several different levels, from the dynein–microtubule interaction during each beating phase to a switch mechanism that turns whole axonemal activities on or off. Here I briefly review what is known about the regulation of flagellar movement.

Waveform Regulation by Calcium

Flagellar waveform of many kinds of organisms are regulated by the calcium ion. A famous example is *Chlamydomonas* flagella, which assume an asymmetric, breast-stroke-like waveform (forward-swimming type) at calcium concentrations lower than 10^{-6} M and a symmetric, sinusoidal waveform (backward-swimming type) at higher Ca^{2+} concentrations. The change in waveform is clearly observed when detergent-extracted flagellar cell models or axonemes are reactivated at different concentrations of Ca^{2+} (Hyams and Borisy, 1978; Bessen et al., 1980). The backward-swimming type of beating is observed when live *Chlamydomonas* cells transiently (up to a few seconds) swim backward upon stimulation by intense light. This action is called the photophobic response. Because cells do not display the photophobic response if the medium is deprived of Ca^{2+} (Schmidt and Eckert, 1976), it is likely that backward swimming in vivo is caused by a Ca^{2+} influx upon light stimulus. Presumably, the light stimulus depolarizes the cell membrane and opens a voltage-sensitive Ca^{2+} channel in the flagellar membrane.

Regulation of the Balance of Beating Between the Two Flagella

Although backward swimming is a very clear cellular response triggered by a light stimulus, it does not appear to be the sole mechanism for tactic behavior in *Chlamydomonas*. This is because analyses of cells' swimming paths indicate that cells gather around bright environments by gradually changing swimming directions, rather than by abruptly reversing course [see, for review, Foster and Smyth, 1980)]. The cell must thus have means to change the swimming direction gradually. This should be achieved by an imbalance between the two flagella of a single cell, because balanced beating of the two flagella, related by an approximately twofold symmetry of the cell, would result only in straight swimming.

Unbalanced beating of the two flagella is observed in detergent-extracted cell models (Kamiya and Witman, 1984). When the models are reactivated with ATP in the absence of Ca^{2+}, many of them circle around in a small area because only one of the two flagella beats; this flagellum is found to be the one nearest the eyespot, a bright membranous organelle located on one side of the cell body. Moreover, when the Ca^{2+} concentration is raised from below 10^{-8} to 10^{-7} M, the balance reverses and the flagellum nearest the eyespot (the cis-flagellum) becomes quiescent while the other (the trans-flagellum) continues to beat. These observations indicate that the two flagella of a Chlamydomonas cell are different and differently controlled by Ca^{2+}. This control may be effected through a gradual modulation of flagellar waveform by minute amounts of Ca^{2+} (Omoto and Brokaw, 1985), or through a change in the beat frequency (Smyth and Berg, 1982). A difference in beat between the two flagella in live Chlamydomonas has been observed by high-speed cinematography by Ruffer and Nultsch (1987); they found that the beating of the two flagella often becomes unbalanced, and it is always the trans-flagellum that tends to have the higher beat frequency. [The intrinsic beat frequency difference between the two flagella has been confirmed by other means (Kamiya and Hasegawa, 1987). Furthermore, it was found that the imbalance is absent in mutants missing the outer dynein arm (Sakakibara and Kamiya, 1989) and enhanced in mutants deficient in the inner arm. These observations point to the importance of outer-arm dynein in producing the cis–trans flagellar imbalance.]

Mechanism of Ca^{2+} Control

The axonemal calcium receptor responsible for the control of waveform remains to be determined. An obvious candidate for the receptor is calmodulin, the presence of which has been demonstrated in flagellar axonemes as well as flagellar membrane (Van Eldik et al., 1980; Gitelman and Witman, 1980). The involvement of calmodulin in motility regulation is unclear, however, because drugs known to disturb calmodulin functions do not, at low concentrations, significantly affect flagellar activity.

Another candidate is the newly found calcium binding protein centrin [or, caltractin (Huang et al., 1988)]. As discussed in detail in Chapter 6, this protein is present in the basal body–nucleus connection (Wright et al., 1985), the flagellar transition region (Sanders and Salisbury, 1989), and one of the inner-arm subspecies in Chlamydomonas (Piperno et al., 1990). Those in the latter two loci may well function in regulating motility.

A further candidate is the calcium-activated phosphatase called calcineurin. It has recently been suggested (Tash, 1989) that a protein similar to calcineurin is associated with outer dynein arms and, in combination with a cAMP-dependent kinase, regulates motility (discussed later) (Tash, 1989). It must, however, be noted that Chlamydomonas mutant axonemes lacking the outer dynein arms still show a Ca^{2+}-dependent behavior such that they stop beating at higher than 10^{-6} M Ca^{2+} (Kamiya and Okamoto, 1985).

This argues against the hypothesis that the outer dynein arm is the sole site for the regulation of Ca^{2+}.

Chlamydomonas mutants that swim backwards only (RL10, RL11; *mbo*1–3) have been isolated (Nakamura, 1979; Segal et al., 1984). Segal et al. (1984) have shown that the *mbo* axonemes are deficient in six proteins and in the beaklike structures that are normally present within certain outer doublets of the wild type (Hoops and Witman, 1983). Because four of the six proteins have been found to be phosphorproteins, Segal et al. have suggested that phosphorylation reactions may be involved in the regulation of waveform. It is not known, however, whether the mutants' backward-type waveform is due to defects in the waveform regulation itself or is simply caused by the waveform being frozen because of serious structural defects within the axoneme unrelated to the regulation mechanism.

It is known that a brief protease treatment of demembranated and reactivated *Chlamydomonas* axonemes often results in loss of the central-pair microtubules and that these axonemes beat in the presence of ATP with the backward-swimming pattern only (Hosokawa and Miki-Noumura, 1987). Although this observation indicates that the central pair is important for the generation of asymmetric waveform (forward-swimming-type beating), it is not clear, for the same reason as given above, whether the central pair is directly involved in the waveform control mechanism.

Regulation of Flagellar Activity
Through Phosphorylation

It has recently been postulated that cyclic nucleotides are as important as Ca^{2+} in the regulation of flagellar motility. A notable example is salmonid sperm, which acquires motility only after some axonemal protein(s) is phosphorylated by an endogeneous cAMP-dependent protein kinase; detergent-extracted cell models of this sperm, obtained before the initiation of swimming, require addition of cAMP and ATP for reactivation (Morisawa and Okuno, 1982). Similar requirements for cAMP have been demonstrated with other kinds of sperm axonemes. Ishiguro et al. (1982) and Tash et al. (1984) have found the target protein for this phosphorylation from sperm axonemes. Tash has demonstrated that a heat-stable target protein named axokinin (MW = 56 kD) is responsible for the cAMP-dependent regulation of motility and that it is present in the flagella and cilia of a variety of organisms, including *Chlamydomonas*. Recent studies indicate that axokinin is probably identical to a regulatory subunit of cAMP-dependent protein kinase (see Tash, 1989).

Although there is evidence that *Chlamydomonas* flagellar activity is affected by the intracellular level of cAMP (Rubin and Filner, 1973), it is not clear whether the motility of this organism is regulated by cAMP: reactivated motility of demembranated flagella does not change when cAMP is added to the reactivation solution (Kamiya, unpublished observation). However, Hasegawa et al. (1987) found that the addition of phosphodiesterase (an enzyme that hydrolyzes cAMP) to the reactivation medium augmented the motility of demembranated flagella; furthermore, an inhibitor of cAMP-dependent kinase also enhanced the reactivated motility of axonemes. These observations suggest that the demembranated flagella of *Chlamydomonas* contain protein kinase and phosphatase activities, as well as an activity to produce cAMP from ATP, and that the motility is negatively controlled by minute amounts of cAMP. The level of cAMP in this organism is shown to increase transiently during the mating reaction (Pasquale and Goodenough, 1987). The inhibition of motility by cAMP may be important for the mating reaction to proceed smoothly.

The cAMP-dependent regulation of flagellar motility, however, may be only a partial aspect of the regulation effected through phosphorylation. It is known that a third of the total of 200 axonemal proteins are phosphorylated in vivo, including the α heavy chain of outer-arm dynein and several heavy chains of the inner-arm dynein (Piperno and Luck, 1981). It is well known that the activities of smooth muscle and nonmuscle myosins are controlled through phosphorylation of their components; by inference, the activity of dynein may also be controlled through phosphorylation. The control of dynein and axonemal activity may be more complicated than that of myosin and myosin–actin systems because, in the former, there are many more components to consider. Multiple kinases–phosphatases may constitute a complicated network in the axoneme. Genetic approaches will therefore be an important means of studying axonemal motility regulation and the basic mechanism for motility generation.

References

Bessen, M., Fay, R. B., and Witman, G. B. 1980. Calcium control of waveform in isolated flagellar axonemes of *Chlamydomonas. J. Cell. Biol.* 80:334–340.

Bloodgood, R. A. 1989. Gliding motility: Can regulated protein movements in the plasma membrane drive whole cell locomotion? *Cell Motil. Cytoskel.* 14:340–344.

Brokaw, C. J. 1982. Models for oscillation and bend propagation by flagella. *Symp. Soc. Exp. Biol.* 35:313–338.

Brokaw, C. J. 1985. Computer simulation of flagellar movement. VI. Simple curvature-controlled models are incompletely specified. *Biophys. J.* 48:633–642.

Brokaw, C. J., and Kimiya, R. 1987. Bending patterns of flagella: IV. Mutants with defects in inner and outer dynein arms indicate differences in dynein arm function. *Cell Motil.* 8:68–75.

Brokaw, C. J., Luck, D. J. L., and Huang, B. P. 1982. Analysis of movement of

Chlamydomonas flagella: The function of the radial spoke system is revealed by comparison of wild type and mutant flagella. *J. Cell Biol.* 92:722–732.

Cooke, R., Crowder, M. S., and Thomas, D. D. 1982. Orientation of spin labels attached to cross-bridges in contracting muscle fibres. *Nature (Lond.)* 300:776–778.

Foltz, K. R., and Asai, D. J. 1990. Molecular cloning and expression of sea urchin embryonic ciliary dynein beta heavy chain. *Cell Motil. Cytoskel.* 16:33–46.

Foster, K. W., and Smyth, R. D. 1980. Light antennas in phototactic algae. *Microbiol. Rev.* 44:572–630.

Gibbons, I. R. 1981. Cilia and flagella of eukaryotes. *J. Cell Biol.* 91:107s–124s.

Gibbons, B. H., Baccetti, B., and Gibbons, I. R. 1985. Live and reactivated motility in the 9 + 0 flagellum of *Anguilla* sperm. *Cell Motil.* 5:333–350.

Gibbons, B. H., and Gibbons, I. R. 1972. Flagellar movement and adenosine triphosphatase activity in sea urchin sperm extracted with Triton X–100. *J. Cell Biol.* 54:75–97.

Gibbons, B. H., and Gibbons, I. R. 1973. The effect of partial extraction of dynein arms on the movement of reactivated sea urchin sperm. *J. Cell Sci.* 13:337–357.

Gibbons, I. R., Shigyoji, C., Murakami, A., and Takahashi, K. 1987. Spontaneous recovery after experimental manipulation of the plane of beat in sperm flagella. *Nature (Lond.)* 325:351–352.

Gitelman, S. E., and Witman, G. B. 1980. Purification of calmodulin from *Chlamydomonas:* Calmodulin occurs in cell bodies and flagella. *J. Cell Biol.* 98:764–770.

Goodenough, U. W., & Heuser, J. E. 1984. Structural comparison of purified dynein proteins with *in situ* dynein arms. *J. Mol. Biol.* 180:1083–1118.

Goodenough, U. W., and Heuser, J. E. 1985. Structure of inner dynein arms, radial spokes, and the central-pair/projection complex of cilia and flagella. *J. Cell Biol.* 100:2008–2018.

Goodenough, U. W., and Heuser, J. E. 1989. Structure of the soluble and in situ ciliary dyneins visualized by quick-freeze deep etch microscopy. In F. D. Warner, P. Satir, and I. R. Gibbons (eds.), *Cell Movement,* Vol. 4. New York, Alan R. Liss, pp. 121–140.

Goodenough, U. W., Gebhart, B., Mermall, V. Mitchell, D. R., and Heuser, J. E. 1987. High-pressure liquid chromatography fractionation of *Chlamydomonas* dynein extracts and characterization of inner-arm dynein subunits. *J. Mol. Biol.* 194:481–494.

Hasegawa, E., Hayashi, H., Asakura, S., and Kamiya, R. 1987. Stimulation of in vitro motility of *Chlamydomonas* by inhibition of cAMP-dependent phosphorylation. *Cell Motil. Cytoskel.* 8:302–311.

Herth, W. 1982. Twist (and rotation?) of central-pair microtubules in flagella of *Poterioochromonas. Protoplasma* 112:17–25.

Hoffman-Berling, H. 1955. Geisselmodelle und Adenosintriphosphat. *Biochem. Biophys. Acta* 16:146–154.

Hoops, H. J., and Witman, G. B. 1983. Outer doublet heterogeneity reveals structural polarity related to beat direction in *Chlamydomonas* flagella. *J. Cell Biol.* 97:902–908.

Hosokawa, Y., and Miki-Noumura, T. 1987. Bending motions of *Chlamydomonas* axonemes after extrusion of central-pair microtubules. *J. Cell Biol.* 105:1297–301.

Huang, B., Piperno, G., and Luck, D. J. L. 1979. Paralyzed flagella mutants of *Chlamydomonas reinhardtii:* Defective for axonemal doublet microtubule arms. *J. Biol. Chem.* 254:3091–3099.

Huang, B., Ramanis, Z., and Luck, D. J. L. 1982. Suppressor mutations in *Chlamydomonas* reveal a regulatory mechanism for flagellar function. *Cell* 28:115–124.

Huang, B., Watterson, D. M., Lee, V. D. & Schibler, M. J. 1988. Purification and characterization of a basal body-associated Ca^{2+}-binding protein. *J. Cell Biol.* 107:121–131.

Hyams, J. S. 1985. Binding of *Tetrahymena* dynein to axonemes of *Marsilea vestita* lacking the outer dynein arm. *J. Cell Sci.* 73:299–310.

Hyams, J. S., and Borisy, G. G. 1978. Isolated flagellar apparatus of *Chlamydomonas*: Characterization of forward swimming and alteration of waveform and reversal of motion by calcium ions *in vitro*. *J. Cell Sci.*, 33:235–253.

Ishiguro, K., Murofushi, H., and Sakai, H. 1982. Evidence that cAMP-dependent protein kinase and a protein factor are involved in reactivation of Triton X–100 models of sea urchin and starfish spermatozoa. *J. Cell Biol.* 92:777–782.

Jarosch, R., and Fuchs, B. 1975. Zur Fibrillenrotation in der *Synura*-Geissel. *Protoplasma* 85:285–290.

Johnson, K. A. 1985. Pathway of the microtubule-dynein ATPase and the structure of dynein: A comparison with actomyosin. *Ann. Rev. Biophys. Biophys. Chem.* 14:161–188.

Johnson, K. A., and Wall, J. S. 1983. Structure and molecular weight of the dynein ATPase. *J. Cell Biol.* 96:669–678.

Kagami, O., Takada, S., and Kamiya, R. 1990. Microtubule translocation caused by three subspecies of inner-arm dynein from *Chlamydomonas* flagella. *FEBS Lett.* 264:179–82.

Kamiya, R. 1982. Extrusion and rotation of the central-pair microtubules in detergent-treated *Chlamydomonas* flagella. *Cell Motil. (Supp.)* 1:169–173.

Kamiya, R. 1988. Mutations at twelve independent loci result in absence of outer dynein arms in *Chlamydomonas reinhardtii*. *J. Cell Biol.* 107:2253–2258.

Kamiya, R., and Hasegawa, E. 1987. Intrinsic difference in beat frequency between the two flagella of *Chlamydomonas reinhardtii*. *Exp. Cell Res.* 173:299–304.

Kamiya, R., Kurimoto, E., and Muto, E. 1991. Two types of *Chlamydomonas* flagellar mutants missing different components of inner-arm dynein. *J. Cell. Biol.*, 112:441–447.

Kamiya, R., Kurimoto, E., Sakakibara, H., and Okagaki, T. 1989. A genetic approach to the function of inner- and outer-arm dynein. In F. D. Warner, P. Satir, and I. R. Gibbons (eds.), *Cell Movement*, Vol. 1. New York, Alan R. Liss, pp. 209–218.

Kamiya, R., and Okagaki, T. 1986. Cyclical bending of two outer-doublet microtubules in frayed axonemes of *Chlamydomonas*. *Cell Motil. Cytoskel.* 6:580–585.

Kamiya, R., and Okamoto, M. 1985. A mutant of *Chlamydomonas reinhardtii* that lacks the flagellar outer dynein arm but can swim. *J. Cell Sci.* 74:181–191.

Kamiya, R., and Witman, G. B. 1984. Submicromolar levels of calcium control the balance of beating between the two flagella in demembranated models of *Chlamydomonas*. *J. Cell Biol.* 98:97–107.

King, S. M., and Witman, G. B. 1989. Molecular structure of *Chlamydomonas* outer arm dynein. In F. D. Warner, P. Satir, and I. R. Gibbons, (eds.), New York, Alan R. Liss, pp. 61–75.

Lee-Eiford, A. Ow, R. A., and Gibbons, I. R. 1986. Specific cleavage of dynein heavy chains by ultraviolet irradiation in the presence of ATP and vanadate. *J. Biol. Chem.* 261:2337–2342.

Lefebvre, P. A., and Rosenbaum, J. L. 1986. Regulation of the synthesis and assembly of ciliary and flagellar proteins during regeneration. *Annu. Rev. Cell Biol.* 2:517–546.

Lewin, R. A. 1952. Studies on the flagella of algae. I. General observations on *Chlamydomonas moewusii* Gerloff. *Biol. Bull.* 103:74–79.

Luck, D. J. L. 1984. Genetic and biochemical dissection of the eukaryotic flagellum. *J. Cell Biol.* 98:789–794.

Melkonian, M., and Preisig, H. R. 1982. Twist of central-pair microtubules in the flagellum of the green flagellate *Scourfieldia caeca*. *Cell Biol. Int. Rep.* 6:269–277.

Mitchell, D. R. 1989. Molecular analysis of the alpha and beta dynein genes of *Chlamydomonas reinhardtii. Cell Motil. Cytoskel.* 14:435–445.

Mitchell, D. R., and Rosenbaum, J. L. 1985. A motile *Chlamydomonas* flagellar mutant that lacks outer dynein arms. *J. Cell Biol.* 100:1228–1234.

Morisawa, M., and Okuno, M. 1982: Cyclic AMP induces maturation of trout sperm axoneme to initiate motility. *Nature (Lond.)* 295:703–704.

Nakamura, S. 1979. A backward swimming mutant of *Chlamydomonas reinhardtii. Exp. Cell Res.* 123:441–444.

Okagaki, T., and Kamiya, R. 1986. Microtubule sliding in mutant *Chlamydomonas* axonemes devoid of outer or inner dynein arms. *J. Cell Biol.* 103:1895–1902.

Omoto, C. K., and Brokaw, C. J. 1985. Bending patterns of *Chlamydomonas* flagella: II. Calcium effects on reactivated *Chlamydomonas* flagella. *Cell Motil.* 5:53–60.

Omoto, C. K., and Johnson, K. 1986. Activation of the dynein adenosinetriphosphatase by microtubules. *Biochemistry* 25:419–427.

Omoto, C. K., and Kung, C. 1980. Rotation and twist of the central-pair microtubules in the cilia of *Paramecium. J. Cell Biol.* 103:1895–1902.

Omoto, C. K., and Witman, G. B. 1981. Functionally significant central-pair rotation in a primitive eukaryotic flagellum. *Nature (Lond.)* 290:708–710.

Paschal, B. M., King, S. M., Moss, A. G., Collins, C. A. Vallee, R. B., and Witman, G. B. 1987a. Isolated flagellar outer arm dynein translocates brain microtubules *in vitro. Nature (Lond.)* 330:672–674.

Paschal, B. M., Shepetner, H. S., and Vallee, R. B. 1987b. MAP1C is a microtubule-activated ATPase which translocates microtubules in vitro and has dynein-like properties. *J. Cell Biol.* 105:1273–1282.

Pasquale, S. M., and Goodenough, U. W. 1987. Cyclic AMP functions as a primary sexual signal in gametes of *Chlamydomonas reinhardtii. J. Cell Biol.* 105:2279–2292.

Pfister, K. K., Fay, R. B., and Witman, G. B. 1982. Purification and polypeptide composition of dynein ATPase from *Chlamydomonas* flagella. *Cell Motil.* 2:525–547.

Piperno, G., and Luck, D. J. L. 1979a. An actin-like protein is a component of axonemes from *Chlamydomonas* flagella. *J. Biol. Chem.* 254:2187–2190.

Piperno, G., and Luck, D. J. L. 1979b. Axonemal adenosine triphosphatases from flagella of *Chlamydomonas reinhardtii*. Purification of two dyneins. *J. Biol. Chem.* 254:3084–3090.

Piperno, G., and Luck, D. J. L. 1981. Inner arm dyneins from flagella of *Chlamydomonas reinhardtii. Cell* 27:331–340.

Piperno, G., Ramanis, Z., Smith, E. F., and Sale, W. S. 1990. Three distinct inner dynein arms in *Chlamydomonas* flagella: Molecular composition and location in the axoneme. *J. Cell Biol.* 110:379–389.

Ringo, D. L. 1967. Flagellar motion and fine structure of the flagellar apparatus in *Chlamydomonas. J. Cell. Biol.* 33:543–571.

Rubin, R. W., and Filner, P. 1973. Adenosine 3',5'-cyclic monophosphate in *Chlamydomonas reinhardtii*. Influence on flagellar function and regeneration. *J. Cell Biol.* 56:628–635.

Ruffer, U., and Nultsch, W. 1987. Comparison of the beating of cis- and trans-flagella of *Chlamydomonas* cells held on micropipettes. *Cell Motil. Cytoskel.* 7:87–93.

Sakakibara, H., and Kamiya, R. 1989. Functional recombination of outer dynein arm-missing flagellar anonemes of a *Chlamydomonas* mutant. *J. Cell Sci.* 92:77–83.

Sale, W. S., and Fox, L. A. 1988. Isolated beta-heavy chain subunit of dynein translocates microtubules in vitro. *J. Cell Biol.* 107:1793–1797.

Sale, W. S., and Satir, P. Direction of active sliding of microtubules in *Tetrahymena* cilia. *Proc. Natl. Acad. Sci. U.S.A.* 74:2045–9.

Sanders, M. A., and Salisbury, J. L. 1989. Centrin-mediated microtubule severing during flagellar excision in *Chlamydomonas reinhardtii*. *J. Cell Biol.* 108:1751–1260.

Satir, P. 1968. Studies on cilia. III. Further studies on the cilium tip and a "sliding filament" model of ciliary motility. *J. Cell Biol.* 39:77–94.

Schmidt, J. A., and Eckert, R. 1976. Calcium couples flagellar reversal to photo stimulation in *Chlamydomonas reinhardtii*. *Nature (Lond.)* 262:713–715.

Segal, R. A., Huang, B., Ramanis, Z., and Luck, D. J. L. 1984. Mutant strains of *Chlamydomonas reinhardtii* that move backwards only. *J. Cell Biol.* 98:2026–2034.

Smyth, R. D., and Berg, H. C. 1982. Change in flagellar beat frequency of *Chlamydomonas* in response to light. *Cell Motil. (Suppl.)* 1:211–215.

Summers, K. E., and Gibbons, I. R. 1971. Adenosine triphosphate-induced sliding of tubules in trypsin-treated flagella of sea urchin sperm. *Proc. Natl. Acad. Sci. U.S.A.* 68:3092–3096.

Tamm, S. L., and Tamm, S. 1981. Ciliary reversal without rotation of axonemal structures in ctenophore comb plates. *J. Cell Biol.* 89:495–509.

Tash, J. S. 1989. Protein phosphorylation: The second messenger signal transducer of flagellar motility. *Cell Motil. Cytoskel.* 14:332–339.

Tash, J. S., Kakar, S. S., and Means, A. R. 1984. Flagellar motility requires the cAMP-dependent phosphorylation of a heat-stable NP–40 soluble 56 kd protein, axokinin. *Cell* 38:551–559.

Vale, R. D., Reese, T. S., and Sheetz, M. P. 1985. Identification of a novel force-generating protein, kinesian, involved in microtubule-based motility. *Cell* 42:39–50.

Vale, R. D., and Yano-Toyoshima, Y. 1988. Rotation and translocation of microtubules *in vitro* induced by dyneins from *Tetrahymena* cilia. *Cell* 52:459–469.

Van Eldik, L. J., Piperno, G., and Watterson, M. D. 1980. Similarities and dissimilarities between calmodulin and a *Chlamydomonas* flagellar protein. *Proc. Natl. Acad. Sci. U.S.A.* 77:4779–4783.

Warner, F. D. 1983. Organization of interdoublet links in *Tetrahymena* cilia. *Cell Motil.* 3:321–332.

Witman, G. B., Plummer, J., and Sander, G. 1978. *Chlamydomonas* flagellar mutants lacking radial spokes and central tubules. Structure, composition, and function of specific axonemal components. *J. Cell Biol.* 76:729–747.

Wright, R. L., Salisbury, J., and Jarvik, J. W. 1985. A nucleus-basal body connector in *Chlamydomonas reinhardtii* that may function in basal body localization or segregation. *J. Cell Biol.* 101:1903–1912.

Yanagida, T., Arata, T., and Oosawa, F. 1985. Sliding distance of actin filament induced by a myosin crossbridge during one ATP hydrolysis cycle. *Nature (Lond.)* 316:366–369.

Yano, Y., and Miki-Noumura, T. 1981. Recovery of sliding ability in arm-depleted flagellar axonemes after recombination with extracted dynein I. *J. Cell Sci.* 48:223–239.

CHAPTER 6

Centrin-Mediated Cell Motility in Algae

Michael Melkonian, Peter L. Beech, Christos Katsaros, and Dorothee Schulze

Historical Perspective: Ciliate Contractility

Centrin-mediated motility represents a novel cell motility mechanism in eukaryotic cells that has attracted considerable attention in recent years (for reviews see Melkonian, 1989; Salisbury, 1989a,b). The principal protein involved in this type of motility, centrin, was first isolated from an algal flagellate (Salisbury et al., 1984) and since then has been found to be a ubiquitous structural protein of the eukaryotic centrosome (Salisbury et al., 1986; Baron and Salisbury, 1988; Hiraoka et al., 1989). Centrin is a Ca^{2+}-modulated phosphoprotein of the EF-hand protein family and shares significant sequence homologies with other members of this family, including calmodulin and the yeast CDC31 gene product (Salisbury et al., 1984; Huang et al., 1988b; Moncrief et al., 1990). Centrin-mediated cell motility is characterized by contraction of a filamentous structure within less than 20 ms, and much slower reextension of this structure to its original length (in the range of a few seconds up to about 1 h). Contraction of centrin filaments is based on supercoiling, not on sliding, initiated by elevated intracellular Ca^{2+} levels and is independent of ATP hydrolysis. Reextension of centrin filaments requires removal of Ca^{2+} from the protein and ATP.

Although cell motility systems with such characteristics have only recently been studied in greater detail, the extraordinarily rapid contractile movements in some of these systems have been a curiosity for several hundred

179

years. A prominent example is the extremely rapid contraction (2–10 ms) of the stalk (spasmoneme) of vorticellid ciliates (van Leeuwenhoek, 1676; Engelmann, 1875; Entz, 1892; Jones et al., 1970). Through the pioneering studies of Hoffmann-Berling (1958) it was recognized that contraction of glycerinated spasmonemes is initiated by low concentrations of Ca^{2+} but is independent of external ATP. Removal of Ca^{2+} by EDTA leads to a reextension of the spasmoneme and repeated cycles of contraction–reextension of the spasmoneme (up to 35 times) could be induced by treating spasmonemes alternately with solutions containing 10^{-8} and 10^{-6} M Ca^{2+} (Hoffmann-Berling, 1958; Amos, 1971) without addition of external ATP and in the presence of various inhibitors of energy metabolism. Hoffmann-Berling (1958) and Amos (1971) also observed that if spasmoneme models were subjected to 2 mM ATP, 4 mM Mg^{2+}, and 10^{-4} M Ca^{2+}, some models exhibited spontaneous cycles of rapid contractions and slow extensions (up to 20 times per minute) resembling the in vivo situation. Since cycles of contraction and extension of the spasmoneme can occur in the absence of metabolic energy by changing only the Ca^{2+} levels, Amos (1971) concluded that the driving force for spasmoneme contraction was the chemical potential of calcium. To estimate whether the change in the chemical potential of calcium would be sufficient to drive spasmoneme contractility, Amos (1971) calculated the amount of work performed by the spasmoneme during a single cycle to be about 11 J kg^{-1} of wet weight. If during spasmoneme contraction there is a change in the chemical potential of calcium between 10^{-6} and 10^{-8} M, then for every mole of Ca^{2+} bound 10^{4} J of energy is available to do work. Thus, in a single contraction cycle, 0.044 g Ca^{2+} kg^{-1} wet weight must be bound to the spasmoneme to make this mechanism possible (Routledge et al., 1976). Using x-ray microanalysis, Routledge et al. (1975) determined that during contraction the spasmoneme bound up to 0.36 g Ca^{2+} kg^{-1} wet weight, more than eight times the amount of Ca^{2+} needed to perform the work as required by the preceding calculation. Ultrastructurally, the spasmoneme consists of 3–4 nm filaments lying parallel to the long axis of the organelle (Amos, 1972, 1975). Dispersed among these filaments are tubular ER elements and mitochondria. The presence of calcium in the tubular ER elements of the spasmoneme in Vorticella and the related Carchesium was demonstrated cytochemically by Carasso and Favard (1966). This led Amos (1971) and Routledge et al. (1976) to propose that Ca^{2+} is released from the tubular ER upon stimulation, binds to the filaments, and leads to spasmoneme contraction, whereas during spasmoneme reextension Ca^{2+} is presumably pumped back into the tubules by a Mg–ATPase system. This hypothesis also explained the cycles of contraction–extension in spasmoneme models in the presence of Ca^{2+} and ATP, assuming that the tubular ER was functionally intact in the glycerinated models.

Further progress in understanding spasmoneme contraction was dependent on isolation and characterization of the Ca^{2+}-binding protein in the spasmoneme. Because of the small quantities of spasmonemes available for analysis (spasmonemes were dissected individually from *Zoothamnion*-colonies!), progress was linked to the development of suitable microanalytical techniques. Amos et al. (1975) finally reported that 60 percent of spasmoneme protein consisted of two similar polypeptides of around 20 kDa, which they termed spasmins (spasmin A and B). No actin or tubulin could be detected by polyacrylamide gel electrophoresis of glycerinated spasmonemes. In the presence of Ca^{2+} the electrophoretic mobility of the spasmins was specifically reduced at the same low Ca^{2+} concentrations that induce spasmoneme contraction (Amos et al., 1975). Whereas spasmins were originally identified in the giant spasmonemes of the colony-forming *Zoothamnion* (Amos et al., 1975), later studies by Routledge (1978) demonstrated that the spasmonemes of *Vorticella* and *Carchesium* also contained a group of closely related proteins of molecular weight around 20 kD. Spasmins are acidic proteins with pI's between 4.7 and 4.8, and they exhibit a high affinity for Ca^{2+} in strong urea solutions, as indicated by an increased electrophoretic mobility in the presence of Ca^{2+} in urea gels (particularly of spasmin B; Routledge 1978). The increase in electrophoretic mobility (anodally) of spasmin B was interpreted by Routledge (1978) to indicate that the binding of Ca^{2+} presumably causes the unfolded protein to become more compact. Amos et al. (1975), in a rough estimation, calculated that an average of 1.4–2.1 Ca^{2+} ions is bound by one molecule of spasmin. Because of the similarities in the characteristics of Ca^{2+} binding between spasmins and some other Ca^{2+}-binding proteins, such as troponin C and parvalbumin, Routledge (1978) suggested that these proteins may have structurally homologous Ca^{2+}-binding sites.

The molecular mechanism of contraction of the spasmoneme filaments remained elusive, however. It was not shown, for example, that the spasmoneme filaments are actually composed of spasmin. Theoretical calculations predicted the diameter of the spasmin molecule to be around 3.6 nm, which is consistent with a longitudinal spacing of 3.5 nm observed in negatively stained spasmoneme filaments (Amos, 1975) and suggested that filaments are linear polymers of spasmin molecules (Amos et al., 1975). Three different hypotheses were advanced to explain the contraction of spasmoneme filaments (Routledge et al., 1976): an electrostatic model (Hoffmann-Berling, 1958), an entropic rubber-band model (Weis-Fogh and Amos, 1972), and a helical, coiled-filament model (Huang and Pitelka, 1973; Kristensen et al., 1974). The latter model was developed to explain another ciliate contractility system, the myoneme contraction of *Stentor* (Huang and Pitelka, 1973). Myonemes are involved in cell body contractions of various heterotrichous ciliates, such as *Condylostoma, Eufolliculina, Spirostomum,* or *Stentor* (Bannister and Tatchell, 1968; Ettienne, 1970; Jones et al., 1970; Huang and

Mazia, 1975). The contractile properties of *Stentor* are characterized by extreme changes in cell body length (to 20–25 percent of its extended length) associated with rapid cell contraction (5 ms) and slow extension. Two cytoskeletal systems contribute to this extraordinary cell contractility (Randall and Jackson, 1958): an overlapping ribbon of microtubules (km fibers) and myonemes (for a review see Pitelka, 1969). Through careful fixations of *Stentor* in the extended and contracted state, Huang and Pitelka (1973) were able to show that the myonemes in the extended state consist of longitudinally oriented dense bundles of 4-nm filaments that transform, upon contraction, to densely aggregated, tubular-appearing 10–12-nm filaments of relatively short lengths. A transformation involving coiling of individual extended filaments upon themselves was envisaged to result in their simultaneous shortening and tubularization (Huang and Pitelka, 1973). Myoneme contraction in *Stentor* and *Spirostomum* exhibits very similar characteristics to spasmoneme contraction: Contraction is dependent on Ca^{2+} but occurs in the absence of ATP (Huang and Mazia, 1975). Recently, Ishida and Shigenaka (1988) succeeded in preparing cell models of *Spirostomum* capable of contraction and extension of the cell body. As in *Stentor,* contraction of the model was induced by increasing the Ca^{2+} concentration in the absence of Mg–ATP; reextension required Mg–ATP without Ca^{2+}. Based on previous ultrastructural investigations, contraction and extension of *Stentor* and *Spirostomum* have been considered to result from antagonistic actions of the myoneme (contraction) and the microtubular ribbon (extension) (Huang and Pitelka, 1973; Etienne and Seltsky, 1974; Yogosawa-Ohara et al., 1985). Huang and Pitelka (1973) proposed an ATP-dependent relative sliding of ribbon microtubules in the reextension of the cell body of *Stentor*. Whether the observed sliding of ribbon microtubules is an active process remains unproven. The ATP requirement for extension could alternatively be explained by Mg–ATPase-dependent pumping of Ca^{2+} from the myoneme into abundant smooth ER that surrounds the myonemes (Huang and Pitelka, 1973)—similar to what has been proposed for spasmoneme extension (see earlier). The inhibition of cell body extension in *Spirostomum* models by vanadate has been interpreted, however, as indicating the involvement of a dyneinlike ATPase in microtubule sliding during cell body extension (Ishida and Shigenaka, 1988).

Although helical coiling of filaments had not been observed in the spasmoneme of contracted vorticellids, Routledge (1978) anticipated that such a mechanism of filament contraction could also operate in the spasmoneme. Both spasmonemes and myonemes undergo a birefringence change upon contraction (Kristensen et al., 1974; Amos, 1971). In the extended state the spasmoneme of *Carchesium* is positively birefringent with respect to its length. On contraction the birefringence falls and the spasmoneme becomes totally isotropic. The observed birefringence is largely form birefringence and not intrinsic birefringence, and thus supports the helical coiled-filament model over the entropic rubber-band model (Routledge et al., 1976). Small changes in the conformation or charge of the spasmin molecule could produce the change in the structure of the spasmin filaments indicated by the birefringence measurements. Routledge (1978) speculated that a change in intersubunit bonding might be produced by a Ca^{2+}-sensitive conformational change in spasmin. Further progress in the analysis of the spasmoneme and myoneme systems of ciliates was presumably hampered by the inability to isolate larger quantities of the filament-forming protein for antibody generation, microsequencing, and analysis of spasmin reassembly. Thus, ciliate contractility remained a somewhat exotic mechanism apparently restricted to a unique group of protists. It took a further 10 years and a new approach using a new organism (the green flagellate *Tetraselmis striata;* see later) to establish spasmin–centrin-mediated motility as a novel and universal cell motility mechanism in eukaryotic cells.

Centrin and the Tetraselmis Rhizoplast

The circumstances that led to the discovery of centrin can best be understood in the context of the numerous descriptive studies performed during the period 1970–1982 on the ultrastructure of the (green) algal flagellar apparatus. These studies were stimulated by the pioneering work of Manton (Manton, 1952, 1956, 1965), who first showed that basal bodies in flagellate algae are usually associated with internal appendages, which she termed *flagellar roots.* Manton (1965) concluded that the variation she found in, for example, flagellar root patterns in various flagellate algae indicated that these structures could be of considerable phyletic significance. Later studies using the chlorophyte flagellar apparatus as an example strongly supported this conclusion (for reviews see Stewart and Mattox, 1975; Pickett-Heaps, 1975; Moestrup, 1978; Mattox and Stewart, 1984; Melkonian, 1984; O'Kelly and Floyd, 1984).

Flagellar roots were first discovered in whole mount preparations of disintegrated, naked cells (Manton, 1952). Only later, when the methodology of osmic acid fixations, plastic embedding, and ultrathin sections became available, were these roots seen in situ. By 1964 it was recognized that most flagellar roots consist of microtubules, which had been recognized as universal constituents of eukaryotic cells only a year earlier through the introduction of the glutaraldehyde fixatives (Sabatini et al., 1963). The greater stability of flagellar root microtubules (and of flagellar microtubules) against fixatives, low temperature, high pressure, and so on, in comparison to cytoplasmic microtubules, was soon recognized. The reason for this property remained obscure until a few years ago, when it was shown that these microtubules are posttranslationally modified (i.e., acetylated; LeDizet and Piperno, 1986). Other flagellar roots are fibrous in nature and, when seen in the electron microscope, appear to connect the basal bodies with the nucleus and thus correspond to the "rhizoplasts" of the early light microscopists (e.g., Dangeard, 1901; Hamburger, 1905; Doflein, 1918; Kater, 1929). One of the most structurally elaborate and massive fibrous flagellar roots is found in the thecate Prasinophyceae (genera *Tetraselmis* and *Scherffelia;* for classification of these taxa see Melkonian, 1990).

Although apparently visualized by light microscopists (Zimmermann, 1925), it was with electron microscopy that the structural complexity of these rhizoplasts was fully revealed (Parke and Manton, 1965). The rhizoplasts of *Tetraselmis* spp. (synonyms: *Platymonas, Prasinocladus*) always occur in numbers of two per cell (for a review see Melkonian, 1980a). Each rhizoplast consists of a bundle of 4–6-nm filaments that are interrupted by crossbars with a spacing of usually more than 160 nm. Rhizoplasts originate proximally at the basal bodies (each rhizoplast branches to link to specific triplets of three adjacent basal bodies; Melkonian and Preisig, 1986), pass along the nuclear surface in opposite directions, and terminate at the plasma membrane about 4–5 μm from their proximal origin. Connections between the rhizoplast and the nuclear envelope and plasma membrane were visualized by thin sections and freeze fracture (Robenek and Melkonian, 1979).

Interest in the compositional and functional significance of the *Tetraselmis* rhizoplast increased greatly when Salisbury and Floyd (1978) reported that the rhizoplast of *Tetraselmis subcordiformis* was a contractile organelle. In their seminal paper the authors showed that cells fixed for electron microscopy in the presence of 0.1–5 mM Ca^{2+} exhibited a contracted rhizoplast that was approximately 65 percent shorter and 30 percent wider than in the extended state (cells fixed in the absence of Ca^{2+}). In the contracted state the cross-banding pattern changed and the repeat distance between two crossbars was significantly less than in the extended state. A cyclic contraction and extension of the rhizoplast was observed in living cells when cells were treated with 2 mM $CaCl_2$ and 5 mM ATP in artificial seawater. In such cells, repeated pulses of inpocketing along the plasma membrane occurred at the sites of rhizoplast attachment. Rhizoplast contraction resulted in gross displacement of the basal apparatus from the flagellar groove toward the cell interior and detachment of the plasma membrane from the theca at the sites of rhizoplast association. Salisbury and Floyd (1978) noted that the observed structural changes in the rhizoplast upon contraction were incompatible with the sliding filament model of actomyosin function. They instead suggested filament assembly–disassembly or filament folding or coiling as alternative mechanisms for contraction. In a discussion of the functional significance of rhizoplast contraction, Salisbury and Floyd (1978) doubted that the observed extreme rhizoplast contractions occurred normally in swimming cells. They suggested that a more subtle contraction prevails that results in a pull on the flagellar apparatus. The authors proposed that the contraction of the rhizoplast could act as a mechanical aid in (1) initiation of the flagellar power and recovery strokes, (2) coordination of the beat cycle, and (3) directional control of the planar beat. The apparent limited occurrence of massive rhizoplasts to *Tetraselmis* was explained by the presumptive geometric constraints imposed on flagellar activity by a deep apical flagellar grove that occurs in *Tetraselmis* and some other taxa of prasinophytes.

Although in the light of present knowledge this functional interpretation of rhizoplast contraction seems out of place (a whole cycle of rhizoplast contraction–extension had to occur with each flagellar beat cycle—i.e., within 20 ms!), this study represented a significant departure from the prevalent view that flagellar roots were functional merely in absorbing mechanical stress imposed on the basal apparatus by the beating flagella. In the following years before mass isolation of rhizoplasts from *Tetraselmis* became possible, two groups continued to investigate the *Tetraselmis* rhizoplast mainly by ultrastructural methods. In an ultrastructural analysis of the freshwater-type species of the genus, *Tetraselmis cordiformis,* Melkonian (1979) noted that, in his fixation, the striation pattern within single rhizoplasts gradually changed from a 105-nm repeat in the proximal parts to a 220-nm repeat in the more distal parts. Rhizoplast contraction thus showed polarity, the proximal parts (close to the basal bodies) being more contracted than the distal parts. A further analysis revealed that the number of contracted crossbands correlated well with the extent of inpocketing of the plasma membrane and the depth of an apical nuclear furrow (Robenek and Melkonian, 1979). These fixations of *T. cordiformis* were carried out in the culture medium and flagella were retained on the cell bodies. In Salisbury and Floyd's experiments, however, because of the "Calcium-shock," flagella were shed and the rhizoplasts were presumably exposed to higher concentrations of Ca^{2+} before chemical fixation, resulting in a complete contraction of all parts of the rhizoplast. Although both types of EM images presumably represent artifacts that follow membrane permeabilization–fixation, the results obtained with *T. cordiformis* showed that rhizoplast contraction could be polar—resulting from localized Ca^{2+} influx and establishment of an internal cytoplasmic Ca^{3+} gradient. Salisbury et al. (1981), in an ultrastructural analysis of the flagellar apparatus of *T. subcordiformis,* described continuities between proximal branches of the rhizoplasts and four compound microtubular–fibrillar flagellar roots that extend anteriorly and anchor the basal apparatus of *Tetraselmis* spp. to the plasma membrane and theca. The four anchoring sites at the plasma membrane represent laminated oval disks, which had earlier been recognized by Manton and Parke (1965) and termed *half-desmosomes* by Schnepf and Maiwald (1970). Salisbury et al. (1981) proposed to replace the term *half-desmosome* by *rhizanchora;* the latter was later modified to the linguistically correct *rhizankyra* (Melkonian and Preisig, 1986). The rhizankyra were proposed to act as shock-absorbing structures that dampen the shear stress generated by alternating flagellar power stroke and rhizoplast contraction (Salisbury et al., 1981). Another possible function for the rhizoplast–rhizankyra complex suggested by Salisbury et al. (1981) was that it might alter the direction of motion of the cells as observed, for example, during photostimulation of cells.

A breakthrough in the experimental analysis of the *Tetraselmis* rhizoplast took place when—in a collaborative effort—mass isolation of rhizoplasts was achieved and the first biochemical characterization of an algal flagellar root protein became possible (Salisbury et al., 1984). The following strategy led to the discovery of centrin: *Tetraselmis striata* was chosen as the test organism because it grows well in synthetic seawater media, the cell division cycle can be synchronized (Ricketts 1979), and a large proportion of cells remain in the flagellate stage throughout growth. An attempt was made to isolate and purify intact single rhizoplasts. Toward this end, thecae were disintegrated by ultrasonication to disrupt linkages between the rhizoplast and the theca (see earlier). After dissolution of membranes with nonionic detergents and separation of larger particles (thecae, starch grains, pyrenoids, etc.) by low-speed centrifugation, the supernatant was subjected to various differential and gradient centrifugations. Rhizoplasts were identified in the various fractions and their enrichment was monitored by whole-mount electron microscopy. Rhizoplasts—especially in their contracted states—could easily be identified in whole-mounts by their triangular shape and distinctive cross-banding pattern. Finally, a protocol using sedimentation in gradients of sucrose and flotation on gradients of colloidal silica, both in the presence of Ca^{2+} (0.1 mM), yielded preparations in which the rhizoplasts constituted up to 70 percent of the total number of particles recognizable in thin sections (Salisbury et al. 1984). When subjected to SDS-PAGE this preparation consisted of 60–65 percent (based on Coomassie Brilliant Blue staining) of a single protein of 20-kD molecular mass, suggesting that it could be the major subunit protein of the rhizoplasts. The protein was further characterized by two-dimensional electrophoresis and shown to consist of two acidic isoforms. The more acidic isoform (beta) is a phosphoprotein (incorporating ^{32}P label in vivo); the less acidic alpha isoform contained no detectable label (Salisbury et al., 1984). In two-dimensional alkaline urea gels both isoforms exhibited a Ca^{2+}-dependent mobility shift that strongly suggested that the 20-kD protein is a Ca^{2+}-binding protein. Finally, a polyclonal antibody monospecific against the 20-kD protein was prepared and the antigen was localized to the rhizoplast of *T. striata* using immunofluorescence and immunogold electron microscopy (Salisbury et al., 1984). Based on these results, Salisbury et al. (1984) suggested that the rhizoplast of *Tetraselmis* consists primarily of a 20-kD Ca^{2+}-modulated contractile protein that they anticipated to be homologous to the spasmins of the ciliate contractile fibers. The rhizoplast protein has a low molecular mass and an acidic isoelectric point, it binds Ca^{2+} with high affinity in the presence of mild denaturants, and it shows Ca^{2+}-induced alterations in molecular conformation. The presence of these characteristics prompted speculation (Salisbury et al., 1984) that the protein might belong to the EF-hand family of Ca^{2+}-binding proteins—whose members include important regulatory proteins such as calmodulin, troponin C, and parvalbumin. This proposal was later verified by sequence comparisons between *Chlamydomonas* centrin and calmodulin from different sources (Huang et al., 1988b). The presence of monospecific antibodies against the 20-kD rhizoplast protein from *T. striata* stimulated a large number of studies searching for homologous antigens in other organisms (summarized in Salisbury, 1989b). An update on the occurrence and possible function of centrin or centrin homologues in the various groups of algae constitutes a major part of this review.

Before we address this topic, however, we return to the *Tetraselmis* rhizoplast and analyze the present status of knowledge regarding its function. Interest in the rhizoplast of *Tetraselmis* diminished when a homologous antigen was discovered in the nucleus–basal body connector ("rhizoplast") of *Chlamydomonas reinhardtii* (Wright et al., 1985). The advantage of using *C. reinhardtii* as the preferred test organism is related to the availability of numerous mutants (including centrin-deficient mutants; Wright et al., 1989), the presence of sexuality, enabling genetic analysis, and the possibility of isolating whole intact cytoskeletons from mutants lacking cell walls. In addition, there is a vast body of literature and knowledge on *C. reinhardtii* (see Harris, 1989). In *Chlamydomonas*, however, the homologous rhizoplast is much less prominent and massive than that in *Tetraselmis;* it terminates distally on the nuclear envelope and does not link to the plasma membrane or cell wall. Thus, the function of the *Chlamydomonas* rhizoplast, at least in interphase cells, must clearly be different from that in *Tetraselmis*.

An analysis of the rhizoplast function in interphase cells of *Tetraselmis* must take into account the special structural properties of the rhizoplast and its association with other cell organelles. An important question here is if, and when, the rhizoplast of *Tetraselmis* contracts in vivo. As in other eukaryotic cells, the cytosolic free Ca^{2+} concentration in the flagellate green algae is presumably also kept at a low and constant level of about 5×10^{-8} M Ca^{2+}. In accordance with this we have previously found that during forward swimming of the naked green flagellate *Spermatozopsis similis* the centrin-containing basal body connecting fiber is relaxed (judged from evaluation of basal body angle), whereas upon photostimulation it contracts (McFadden et al., 1987). Experiments with isolated cytoskeletons showed that a rise in the Ca^{2+} concentration above 5×10^{-8} M Ca^{2+} elicited a similar response of the connecting fiber, suggesting that during normal forward swimming of green flagellates the Ca^{2+} concentration is below 10^{-7} M Ca^{2+} and centrin-containing organelles are nonmotile. Results with similar threshold levels of Ca^{2+} have recently been obtained in relation to contractility of the nucleus–basal body connector of *Dunaliella bioculata* (Merten and Melkonian, unpublished observations). If these data can be transferred to *Tetraselmis,* the rhizoplast should be nonmotile in nonstimulated cells. Evidence has been presented that in *Scherffelia dubia,* an organism closely related to *Tetraselmis* spp. (Melkonian and Preisig, 1986), in vivo contraction of the rhizoplast occurs upon flagellar shedding (McFadden and Melkonian, 1986). Flagellar shedding regularly occurs in thecate Prasinophyceae before cell division (Reize and Melkonian, 1987). Flagellar shedding can be experimentally induced in *Tetraselmis* and *Scherffelia* by pH shock or mechanical shearing in the presence of millimolar concentrations of external Ca^{2+} (McFadden and Melkonian, 1986; Reize and Melkonian, 1987; Martindale and Salisbury, 1990), by "calcium-shock" (Salisbury et al., 1984; but see contrary results by Martindale and Salisbury 1990), or by treatment with ethanol (Martindale and Salisbury, 1990). In *Scherffelia dubia,* rhizoplast contraction upon flagellar shedding is polar; the more contracted cross-bands are located close to the flagellar bases (McFadden and Melkonian, 1986). This suggests that Ca^{2+} influx into the cell near the flagellar bases is necessary for rhizoplast contraction. Unfortunately, Martindale and Salisbury (1990) did not take measures to preclude Ca^{2+} entry into their cells from the incubation medium (1 mM Ca^{2+}) during pipetting, centrifugation, and glutaraldeyde fixation following the experimental treatments (pH shock, ethanol, heat shock, A23187 ionophore, "Ca^{2+} shock")—thus making interpretation of the effects of the various treatments on flagellar shedding and rhizoplast contraction difficult.

Concomitant with flagellar shedding and rhizoplast contraction in *Tetraselmis subcordiformis* and *Scherffelia dubia,* the thecal slit from which the flagella normally emerge closes over (Lewin and Lee, 1985; McFadden and Melkonian, 1986; Melkonian and Preisig, 1986). We propose that the closure of the thecal slit is directly caused by rhizoplast contraction mediated through the four rhizankyra, which are optimally positioned at the base of the flagellar grove along the long sides of the thecal slit (Melkonian and Preisig, 1986). Theoretically, closure of the thecal slit could cause flagellar abscission; however, an ultrastructural analysis of early stages of flagellar abscission in *Tetraselmis subcordiformis* by Lewin and Lee (1985) indicates that the two processes are independent—breakage of axonemal doublets occurs well before the thecal slit impinges on the flagella. In addition, flagellar abscission always occurs near the distal end of the stellate structure and the plasma membrane closes transversely over the stump (Lewin and Lee, 1985; Melkonian, unpublished observations). Centrin is apparently present in the flagellar transitional region of *Scherffelia dubia* (unpublished observations), which may indicate that a centrin-related mechanism could be involved in flagellar shedding (see also Sanders and Salisbury, 1989).

The rhizankyra remain attached to the theca following experimental flagellar shedding, and after about 10–20 minutes the regenerating flagellar stumps begin to separate and penetrate the thecal slit. At this stage the rhizoplasts are, surprisingly, still largely contracted (McFadden and Melkonian, 1986). They remain contracted for at least 60 min after flagellar shedding and then slowly reextend to become fully extended only when flagellar regeneration is complete (McFadden and Melkonian, 1986). This suggests that the outgrowing flagella possibly exert considerable force on the thecal slit. Since cells with short regenerating flagella are capable of normal swimming, and rhizoplasts are largely contracted during this time, changes in rhizoplast contractility are presumably not involved in regulating flagellar motility (discussed earlier). The shedding of flagella that signals the onset of mitosis differs in at least one important aspect from experimental flagellar shedding, namely, that the protoplast, including the attachment points of the rhizoplast–rhizankyra complex with the plasma membrane, dissociates completely from the theca. In this case the thecal slit is permanently closed, and following cell division, the zoospores emerge from the parental theca not through the thecal slit, but through rupture of the parental theca in an unspecified manner.

What is the functional significance of thecal closure? *Tetraselmis* and *Scherffelia* differ from all other green flagellates in that they assemble a cell wall (i.e., the theca), composed of highly acidic polysaccharides, by extracellular Ca^{2+}-mediated fusion and coalescence of Golgi-derived scales (for review see Melkonian et al., 1991). Similar scales also form several layers on the flagellar surface of these taxa, and at least the outer two layers of scales presumably assemble from the outside to the flagellar surface. It is therefore conceivable that the closed theca provides an optimal ionic microenvironment for scale assembly and prevents loss of scales to the medium.

In addition to the preceding proposed specialized function of the *Tetraselmis* rhizoplast during interphase, there are certainly other more general functions of the rhizoplast that we will discuss in the later section on the nucleus–basal body connector of green algae.

Centrin: Molecular Mechanism of Its Function

Although centrin has been found in almost all eukaryotic cells that have been investigated so far, important aspects of its molecular function are still unresolved (reviews: Salisbury 1989a,b; Melkonian, 1989). The term *centrin* was introduced by Salisbury and collaborators in two abstracts published in 1987 (Coling and Salisbury, 1987; Baron and Salisbury, 1987). In one abstract it was referred to as the 20-kD rhizoplast protein of *Tetraselmis striata,* in the other, as an antigenically related protein of high molecular mass (165 kD) from mammalian PTK_2 cells. Adding to the confusion, Huang and colleagues purified and characterized a 20-kD Ca^{2+}-binding protein from the basal apparatus of *Chlamydomonas reinhardtii* and named it caltractin (Huang et al., 1988b)—a protein presumably identical to centrin from *Tetraselmis.* The 165-kD protein of PTK_2 cells is now referred to as "centrin-related" (Baron et al., 1991), while a 19.5-kD protein has recently been identified in avian centrosomes and could be a centrin homologue (Salisbury and Greenwood, 1990). Since the term *centrin* is now widely used for a 20-kD Ca^{2+}-modulated EF-hand protein located in contractile fibrous flagellar roots of algae, we wish to retain this name but propose to restrict it to low-molecular-weight proteins sharing the preceding properties with algal centrin.

When centrin was first isolated from the *Tetraselmis* rhizoplast, it was suggested that the phosphorylation–dephosphorylation status of the protein was related to rhizoplast contractility (Salisbury et al., 1984). Comparison of relative amounts of the two isoforms illustrated a variable stoichiometry that ranged from 2:1 to 1:1 (alpha/beta). Preparations from contracted rhizoplasts showed a higher ratio of alpha to beta. It was thus proposed that the beta form is dephosphorylated during rhizoplast contraction and phosphorylated during rhizoplast extension (Salisbury et al., 1984). Rhizoplast contraction had previously been shown to involve a twisting and supercoiling of individual filaments upon exposure to elevated levels of Ca^{2+}, indicating that it is driven by a Ca^{2+}-induced conformational change in filament substructure (Salisbury, 1983). Calcium cytochemistry using potassium pyroantimonate and x-ray microanalysis suggested that *Tetraselmis subcordiformis* contains Ca^{2+}-sequestering vesicles in the vicinity of the rhizoplasts that were postulated to be involved in regulating Ca^{2+}-induced contraction of rhizoplasts (Salisbury, 1982). In addition, calcium–antimonate precipitates were observed in the contracted portions of the rhizoplast but were absent from extended rhizoplasts (Salisbury, 1982). In a preliminary report, Salisbury (1983) also published results on cytochemical localization of ATPase that showed the reaction product (lead phosphate precipitate) to be localized on the cross-striations of the rhizoplast as well as near the limiting membranes of the Ca^{2+}-sequestering vesicles. In summary, these data were interpreted to suggest two energy-requiring steps in rhizoplast contractility: one involved in regulation of cytosolic free Ca^{2+} levels and a second more directly involved in centrin phosphorylation (Salisbury et al., 1984). Whereas the importance of Ca^{2+} for rhizoplast contractility and centrin function in general appears to be firmly established, the significance of centrin phosphorylation–dephosphorylation for centrin-mediated cell motility is much less clear. Two in vitro models for analyzing centrin function have been established in recent years that allow us to address some of these questions. Nucleus–basal body complexes have been isolated from cell wall–less mutants of *Chlamydomonas reinhardtii* by treatment with nonionic detergents (Wright et al., 1985). They consist of the basal apparatus, including two centrin-containing fibers, and the nuclear skeleton to which the basal apparatus remained attached through the centrin fibers (nucleus–basal body connectors, rhizoplasts). Calcium treatment (millimolar Ca^{2+} concentrations) of the complexes caused a shortening of the nucleus–basal body connectors (NBBC) and hence movement of the nuclear skeleton and nuclear DNA toward the basal apparatus (Wright et al., 1985). The average distance between the basal apparatus and the nuclear skeleton decreased from 1.2 μm in the absence of free Ca^{2+} to about 0.034 μm after addition of 1 mM Ca^{2+}. A similar movement of the nucleus apparently also occurs in vivo in *C. reinhardtii* upon flagellar shedding induced by pH shock or mechanical shear (Salisbury et al., 1987). Evaluation of nuclear movement, however, was not done on live cells but followed chemical fixation by immunofluorescence and DAPI staining (Salisbury et al., 1987). Under these conditions it was observed that contraction of the NBBC is very rapid (less than 1 s), but that reextension requires considerably more time (60 min) and parallels flagellar regeneration in a manner similar to that observed earlier in the thecate Prasinophyceae (discussed earlier). Contraction of the NBBC upon flagellar shedding in *C. reinhardtii* requires external Ca^{2+}, since at low external Ca^{2+} concentrations (10^{-7} M Ca^{2+}) no movement of the nucleus occurred upon flagellar shedding and flagellar regrowth was minimal (Salisbury et al., 1987). Unfortunately, these studies did not determine what intracellular Ca^{2+} concentration is needed to initiate contraction of the NBBC. Toward this end we have recently shown that in isolated nucleus–basal body complexes of *Dunaliella bioculata*, nuclear movement is initiated at $>5 \times 10^{-8}$ M Ca^{2+} (Merten and Melkonian, unpublished observations; see also Figs. 6.35, 6.36).

In 1987 McFadden et al. developed a second in vitro system for the study of centrin function: isolated cytoskeletons of the naked green flagellate *Spermatozopsis similis* (McFadden et al., 1987). *Spermatozopsis similis* resembles *Chlamydomonas* in general and cytoskeletal ultrastructure (Preisig and Melkonian, 1984; Melkonian and Preisig, 1984; Lechtreck et al., 1989). In contrast to *C. reinhardtii*, it is naturally wall-less, which greatly facilitates isolation of intact cytoskeletons by detergent treatment (McFadden et al., 1987). During forward swimming the two basal bodies are held in antiparallel position (a 180° angle between them), whereas during backward swimming they are reoriented to lie parallel to each other. Backward swimming (with an undulatory type of beat of the two flagella) is initiated as an avoidance reaction upon photostimulation (i.e., the photophobic response) and lasts only a few seconds before cells resume forward swimming (with an asymmetric "breast stroke"–type of flagellar beat). Basal body reorientation can be evaluated by light microscopy of live cells (following immobilization of cells in agarose; see Reize and Melkonian, 1989), in chemically fixed cells (fixed in the presence of EGTA to avoid Ca^{2+} entry upon fixation), and in isolated cytoskeletons. In live cells it was found that at external Ca^{2+} concentrations greater than 10^{-5} M, photostimulation induces flagellar reversal and basal body reorientation (McFadden et al., 1987). At Ca^{2+} concentrations of less than 10^{-5} M, cells did not react to photostimulation with a photophobic response. Although the cells resumed forward swimming a few seconds after the normal photophobic response, it was found that restoration of the antiparallel basal body configuration did not occur concomitantly, but only 3–4 minutes later (McFadden et al., 1987). Isolated cytoskeletons from S. similis could be reactivated in vitro. Both flagellar beat form and basal body orientation were dependent on the Ca^{2+} concentration; the threshold concentration for basal body reorientation was found to be around 5×10^{-8} M Ca^{2+}, slightly lower than the threshold for flagellar reversal (10^{-7} M Ca^{2+}). By changing the Ca^{2+} concentration between two values (10^{-6} M and 10^{-9} M) it was possible to switch flagellar beat form repeatedly from symmetrical (cytoskeletons backward swimming) to asymmetrical (cytoskeletons forward swimming). Basal body reorientation in such cytoskeletons occurred only once, however, and stayed in the parallel configuration irrespective of further switches in the Ca^{2+} concentration (McFadden et al., 1987). McFadden et al. (1987) presented evidence that the motor for basal body reorientation resides in a centrin-containing fiber that interconnects the two basal bodies near their distal end (distal connecting fiber) and that the motor is principally independent of flagellar beating. The connecting fiber apparently contracts when the Ca^{2+} concentration increases. The reason why the fiber remains in the contracted state when the Ca^{2+} concentration is again lowered is unclear, but some factor critical to the mechanism of distal fiber extension may have been lost or inactivated in the in vitro system. In this regard it is interesting to note that it has also not been possible to achieve NBBC reextension in isolated cytoskeletons of Dunaliella bioculata after lowering the Ca^{2+} concentration to 10^{-9} M (up to 20 h exposure), indicating that this system behaves similarly to the connecting fiber of Spermatozopsis (Merten and Melkonian, unpublished observations). Somewhat surprisingly, Salisbury et al. (1987) found that cell models of wild-type Chlamydomonas reinhardtii obtained by treatment of cells with very low detergent concentrations (0.01 percent NP–40 in buffers containing 3 mM EGTA), when transferred from high Ca^{2+} concentrations (3 mM $CaCl_2$) back to low Ca^{2+} concentrations (3 mM EGTA), revealed a reextension of the NBBC and repositioning of the nucleus to a central position. No NBBC contraction occurred, however, if the models were returned for a second exposure to high Ca^{2+} conditions. But if the second exposure to high Ca^{2+} was done in the presence of 3 mM ATP, NBBC contraction and nuclear movement were again observed (Salisbury et al., 1987). Cell models of C. reinhardtii previously obtained by treatment with much higher detergent concentrations (0.25–0.5 percent NP–40) apparently retained all their internal membranes and an energy metabolism comparable to that of control cells but lost the flagellar membrane from the transitional region outward (Goodenough, 1983). Goodenough (1983) found it difficult to "state whether the extracted Chlamydomonas cells are 'dead' or 'alive'." It is thus not clear to what extent repeated NBBC contractions and reextensions observed in Chlamydomonas "models" obtained by even significantly lower detergent concentrations relate to an in vitro or in vivo condition. Salisbury et al. (1987) interpreted their results to indicate that "phosphorylation is a necessary step in the potentiation of the flagellar root protein for subsequent calcium-induced contraction." Since, however, recent data obtained on rhizoplast contractility in Tetraselmis striata indicate that under certain conditions (heat shock, ethanol treatment) phosphorylation–dephosphorylation of centrin is uncoupled from rhizoplast extension–contraction (Martindale and Salisbury, 1990), the importance of centrin phosphorylation–dephosphorylation for centrin-mediated cell motility is questionable. Other possible roles for protein phosphorylation in fibrous flagellar roots of green algae have been discussed recently by Lechtreck and Melkonian (1991).

A significant advance toward an understanding of the molecular mechanism of centrin function was made when a cDNA was isolated and characterized that contained the entire coding sequence of centrin (synonym: caltractin) from Chlamydomonas reinhardtii (Huang et al., 1988b). Additionally, preliminary DNA and RNA blot analysis indicated that centrin is encoded by a single-copy gene; a single-size transcript of about 1.1 kb was detected (Huang et al., 1988b). The deduced amino acid sequence of the protein showed strong sequence relatedness (45–48 percent identity) with calmodulin from Chlamydomonas and even stronger relatedness (50 percent identity) with the yeast CDC31 gene product required for spindle pole duplication (Huang et al., 1988b). The primary protein sequence of Chlamydomonas centrin and its predicted secondary structure suggests that the protein contains four Ca^{2+}-binding domains of the helix–loop–helix (EF-hand) structure also found in calmodulin and related Ca^{2+}-modulated proteins. The open reading frame of the sequenced cDNA encodes a polypeptide of 169 amino acids with a calculated molecular mass of 19,459 D. Analysis of the predicted amino acid sequence identified two consensus sequences for possible post-translational modifications of centrin: One is a potential c-AMP-dependent phosphorylation site at serine–167; the other, at Asp–133, is a possible attachment site of N-linked oligosaccharides (Huang et al., 1988b). Although it is known that centrin is a phosphoprotein, currently nothing suggests that centrin is, in situ, a glycoprotein. The major difference between calmodulin and centrin is an N-terminal domain in centrin of 21 amino acids that is not found in calmodulin (Huang et al., 1988b). This domain contains a concentration of positively charged amino acids and contains the only tyrosine residue of the protein. A computer search for sequence similarities with this domain revealed no homologous sequences in other known proteins (Sensen and Melkonian, unpublished). A further search for potential epitopes throughout the molecule indicated that at least two possible epitopes occurred in this N-terminal domain, suggesting that the monospecific antibodies available against centrin might recognize one or both of these potential epitopes. In this context, we note that anticentrin antibodies do not cross-react with calmodulin (Huang et al., 1988a). Epitope mapping of centrin, using sequence-specific proteases, has not yet been performed, to our knowledge. It is likely that the N-terminal domain of centrin confers the cytoskeletal properties to the protein; it may be involved in filament formation.

Understanding the molecular function of centrin would be greatly facilitated if it were possible to reassemble in vitro filaments containing centrin from the monomer molecules. Preliminary observations reported by Salisbury et al. (1984) suggest that urea-extracted rhizoplasts from *Tetraselmis striata* can be reassembled into a meshwork of 3-nm filaments upon dialysis of urea. This approach, perhaps in combination with in vitro expression and mutagenesis of centrin, should significantly help to unravel the molecular mechanism of centrin function. At present, the following questions remain unsolved: Is centrin really a filament forming protein and is it sufficient for filament formation? How is Ca^{2+} binding to centrin transformed to twisting and supercoiling of centrin-containing filaments? What is the role of centrin phosphorylation–dephosphorylation? Does it modulate Ca^{2+} binding to centrin? What are the requirements for reextension of centrin-containing filaments? How is Ca^{2+}-removed from the protein?

Occurrence of Centrin and Centrin-Mediated Cell Motility in Different Groups of Algae

In this chapter we review the known occurrence of centrin in the various groups of algae. This is based on immunolocalization and in some cases immunoblot analysis. We identify centrin-containing structures at the light and electron-microscope level and refer to the possible function of centrin-containing structures as they relate to cell motility.

Chlorophyta

As discussed in detail in the preceding chapters, centrin was originally identified in the thecate prasinophyte *Tetraselmis striata*, where it constitutes the principal component of the two massive rhizoplasts of the cell (Salisbury et al., 1984). These structures belong to a type of fibrous flagellar roots known as system II fibers (for reviews see Melkonian, 1980a; Lechtreck and Melkonian, 1991). Schulze et al. (1987) have surveyed 28 taxa of flagellate green algae for the presence of centrin in system II fibers. System II fibers immunoreactive to anticentrin were found in all taxa, except spermatozoids of *Oedogonium cardiacum* and the colorless chlamydomonad *Polytomella parva*. They are most conspicuous in the prasinophyte taxa investigated (see also Figs. 6.1–6.3, 6.11, 6.13) but have also been found in representatives from other green algal classes (i.e., Ulvophyceae, Chlorophyceae, and Pleurastrophyceae; Schulze et al., 1987). System II fibers provide a physical linkage between the flagellar basal apparatus and the nucleus and thus enable nucleus–basal apparatus complexes to be isolated from many flagellate green algal taxa, especially those that lack cell walls or a scaly cell covering (e.g., *Dunaliella bioculata*, Figs. 6.35, 6.36 or *Micromonas pusilla*, Fig. 6.6). Using the sensitive immunofluorescence method, system II fibers (or nucleus–basal body connectors) have also been found in taxa in which they have previously not been detected with conventional electron microscopy. [e.g., *Spermatozopsis similis* (Melkonian and Preisig, 1984), *Hafniomonas reticulata* Ettl and Mosestrup, 1980)]. Using the "preembedding method" of immunogold electron microscopy, we have also been able, however, to visualize the system II fibers clearly in these taxa at the EM-level (Figs. 6.14–6.17). In addition to determining the number of system II fibers per cell in the various taxa (1–4), it was informative to analyze their point of termination (Schulze et al., 1987). The system II fibers may terminate at the nucleus or pass over the nucleus and terminate elsewhere in the cell. The first pattern was found in the Chlorophyceae and Pleurastrophyceae, the second pattern in the Prasinophyceae and Ulvophyceae. In the athecate Prasinophyceae (for a circumscription of the Prasinophyceae see Melkonian, 1990) system II fibers terminate at the chloroplast, often near the pyrenoid (Fig. 6.2), whereas in the thecate Prasinophyceae (Chlorodendrales sensu Melkonian, 1990) the two massive system II fibers (rhizoplasts) terminate at the plasma membrane (see earlier; Fig. 6.1). Also in the ulvophycean taxa studied to date, system II fibers link with the plasma membrane (Melkonian et al., 1988, and Fig. 6.12). In the latter taxa, the association between the basal apparatus and the nucleus was often not stable, and intact nucleus–basal body complexes could not be isolated (Schulze et al., 1987). Schulze et al. (1987) concluded from their study that fibrous flagellar roots (of the system II fiber type) occur universally in flagellate green algal cells and always link the basal bodies

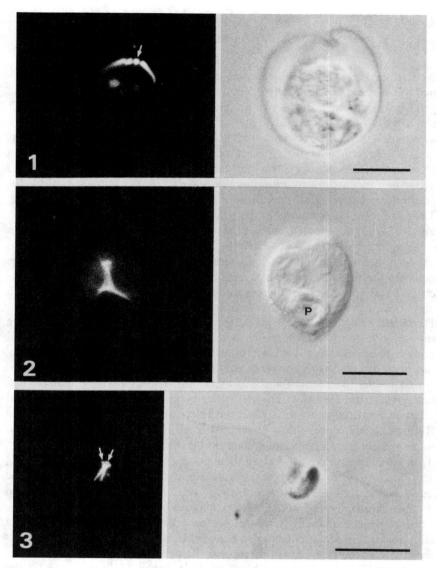

Figures 6.1 to 6.3. Anticentrin immunofluorescence and corresponding phase contrast images of permeabilized, prasinophyte cells. Bar = 10 μm.

(1) *Tetraselmis cordiformis.* Two system II fibers are antigenic and descend from the region of the basal bodies where distal connecting fibers are also labeled (arrow).

(2) *Pyramimonas amylifera.* The numerous system II fibers appear as a single, antigenic band that descends from the basal body region and spreads over the anterior surface of the pyrenoid (P).

(3) *Nephroselmis rotunda.* One branching system II fiber, or two separate fibers, are labeled in addition to basal body conecting fibers (arrows).

Cultures and methods of cell permeabilization and immunoflurescence are as described in Schulze et al. (1987).

with the nucleus. Using immunofluorescence, immunoelectron microscopy, and immunoblotting, it was shown that these system II fibers consist at least in part of centrin (Schulze et al., 1987).

Several possible functions for system II fibers of green algae have been suggested: an essential role in basal body localization and segregation (Wright et al., 1985; Wright et al., 1989); a role in the redetermination of cell polarity through an alteration of the orientation of centrosomal microtubule organizing centers (Salisbury et al., 1986); a role in initiation, coordination, and directional control of flagellar beat (Salisbury and Floyd, 1978; Salisbury et al., 1981; White and Brown, 1981; Rüffer and Nultsch, 1985; see discussion on the function of the *Tetraselmis* rhizoplast); a role in gametic fusion (Melkonian, 1980b); a role in the transduction of signals for flagellar resorption or for induction of protein synthesis following deflagellation (Salisbury et al., 1987); and a role in the poleward movement of chromosomes during anaphase in mitotic cells (Salisbury et al., 1988). From the discussion in the preceding chapters it is clear that in unstimulated cells system II fibers do not change length. Stimulation leads to an increase in Ca^{2+} (presumably by Ca^{2+} influx from the external medium) near the system II fibers and consequently to system II fiber contraction. Contraction is extremely rapid and in some experiments the fiber contracts to less than 50 percent of its original length. Whether such an extraordinary contraction of system II fibers is artifactual or not is not entirely clear. One result of system II fiber contraction is that the nucleus moves up to the basal apparatus (Salisbury et al., 1987). What is the significance of nuclear movement upon contraction of system II fibers? Nuclear movement occurs either at the onset of mitosis, especially when this process is coupled to flagellar retraction or shedding (Salisbury et al., 1988), or during accidental loss of flagella followed by flagellar regeneration (see earlier). In the first case the close association of basal bodies and nucleus might ensure correct duplication and segregation of the basal bodies during mitosis (Wright et al., 1989). In the latter case it has been suggested that system II fiber contraction might be causally related to the signaling of induction of flagellar precursor genes or to the transport of flagellar precursors or their messages to sites of synthesis or assembly near the basal bodies (Salisbury et al., 1987). In their study of the vfl–2 mutant of *Chlamydomonas reinhardtii,* which is largely devoid of centrin (especially in the NBBC region), Wright et al. (1989) showed that deflagellation induced flagellar regeneration with the same kinetics as in wild-type cells. In addition, the level of α tubulin-mRNA increased to the same extent following deflagellation in vfl–2 and wild-type cells (Wright et al., 1989). The results indicate that system II fiber contraction plays no role in gene activation following deflagellation. These results, however, do not preclude the possibility that the same signal that leads to system II fiber contraction (i.e., Ca^{2+} influx) also activates flagella-specific genes directly or through a signaling cascade.

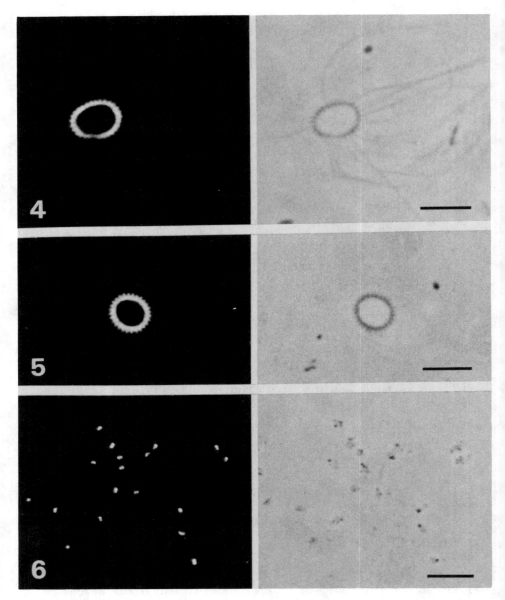

Figures 6.4 to 6.6. Anticentrin immunofluorescence and corresponding phase contrast images of isolated, green algal cytoskeletons. Bar = 5 μm.

(4, 5) Male gametes of *Oedogonium cardiacum* with flagella attached (4) and detached (5). A bright, antigenic ring of distal connecting fibers is visible.

(6) *Micromonas pusilla*. Two antigenic points are associated with the basal body region of each cytoskeleton: one point probably represents the distal connecting fiber, the other the system II fiber.

Cultures and methods of cytoskeleton isolation and immunofluorescence are as described in Schulze et al. (1987).

Since reextension of system II fibers following deflagellation is relatively slow and proceeds only in parallel with flagellar regrowth (McFadden and Melkonian, 1987; Salisbury et al., 1987; and own unpublished observations), contraction of system II fibers upon deflagellation might serve another function: system II fibers could simply act as Ca^{2+} buffers, reducing the free Ca^{2+} concentration to low levels, thus facilitating assembly of tubulin and other flagellar precursors. In accordance with this we note that in taxa living at higher ambient Ca^{2+} concentrations (e.g., in seawater) system II fibers are usually better developed than in taxa exposed to lower Ca^{2+} concentrations (compare Prasinophyceae and Ulvophyceae with Chlorophyceae).

In addition to system II fibers, centrin has been found in flagellate green algae at two other locations within the basal apparatus: in the distal connecting fiber and the flagellar transitional region.

It appears that the location of centrin in the distal connecting fiber is another universal attribute of flagellate green algae. Centrin was first found in the distal connecting fiber of *Spermatozopsis similis*, where it was shown to be involved in basal body reorientation during the photophobic stop response (McFadden et al., 1987; see earlier). Subsequent studies have demonstrated centrin in the distal connecting fibers of numerous other green algal taxa, including *Chlamydomonas reinhardtii* (Schulze et al., 1987; Salisbury et al., 1987; Melkonian et al., 1988; Salisbury et al., 1988; see also Figs. 6.4, 6.5, 6.8–6.12, 6.14–6.19). One of the most spectacular images of centrin localization in algae was obtained from the stephanokont spermatozoids of *Oedogonium cardiacum* (Figs. 6.4, 6.5). Here basal bodies are interconnected by a fibrous ring representing a multiple distal connecting fiber (Pickett-Heaps, 1971). The ring contains regularly spaced fluorescent spikes on its outer surface, their number corresponding to the number of basal bodies present in the basal apparatus (Fig. 6.5). Rings are contractile if exposed to elevated Ca^{2+} concentrations, and their diameter changes in consequence (unpublished observations). Similar contractions of anterior ends of live spermatozoids were described by Hoffman (1973) upon penetration of sperm cells through the oogonial pore.

Centrin is apparently not always homogeneously distributed within the distal connecting fiber (unlike the situation in the system II fibers as presently known). There is considerable variability in the detailed structure of distal connecting fibers in flagellate green algae (review by Melkonian, 1980a). Some fibers contain prominent cross-bands (e.g., in *Chlamydomonas reinhardtii*; Ringo, 1967). In these fibers centrin labeling is presumably restricted to specific cross-bands (Schulze et al., 1987; Melkonian et al., 1988; Salisbury et al., 1988; Figs. 6.10, 6.11). In species of *Spermatozopsis*, however, no cross-bands are observed and centrin distribution is homogeneous throughout the fiber (McFadden et al., 1987; Schulze et al., 1987; Melkonian et al., 1988; Fig. 6.14). As far as we know, centrin does not occur in any

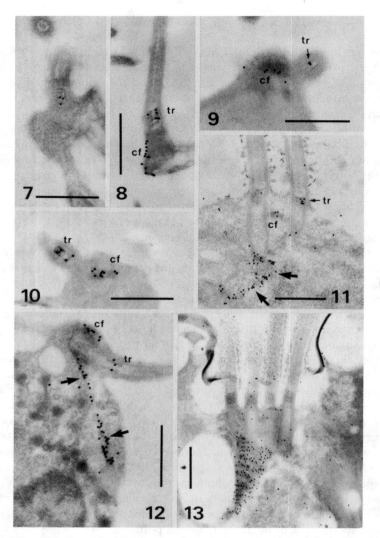

Figures 6.7 to 6.13. Postembedding immunoelectron microscopy of green algal cells embedded in Lowicryl K4M resin and labeled with anti-centrin and protein A/gold. Bar = 0.5 μm.

(7) *Pedinomonas tuberculata*. The flagellar transitional region is labeled.

(8) *Spermatozopsis similis*. The flagellar transitional region (tr) and the distal connecting fiber (cf) are labeled.

(9) *Dunaliellla bioculata*. Labeled as for (8).

(10) *Chlamydomonas reinhardtii* CW 15⁺. Labeled as for (8), but note that the distal connecting fiber shows a heterogeneous antigenicity.

(11) *Pyramimonas obovata*. Labeling is principally on the flagellar transitional region (tr), distal connecting fiber (cf: heterogeneously antigenic), and system II fibers (arrows).

(12) Gametes of *Enteromorpha* sp. Labeling as for Fig. 11.

(13) *Scherffelia dubia*. The cross-striated system II fibers are heavily labeled.

Cultures and methods of fixation and immunolabeling are as described in Schulze et al. (1987).

other type of connecting fiber in flagellate green algae, especially not in the proximal connecting fibers (Beech, unpublished observations).

What is the function of centrin in distal connecting fibers? Contraction of the distal connecting fiber in *Spermatozopsis similis* leads to basal body reorientation (McFadden et al., 1987). We have evidence that similar, albeit not as extensive, basal body reorientations occur in other flagellate green algae [e.g., in *Dunaliella bioculata, Nephroselmis olivacea,* and zoospores of different taxa (unpublished observations)]. In most scaly green flagellates, however, basal bodies are oriented parallel, irrespective of the direction of swimming, and centrin also occurs in the distal connecting fibers (Schulze et al., 1987; Fib. 6.11). The same holds true for the peculiar green flagellate *Scourfieldia caeca* (Fig. 6.19). In addition, no evidence has been obtained so far that pronounced basal body reorientation occurs in species of *Chlamydomonas* (Ringo, 1967; Hyams and Borisy, 1978). Reorientation of the basal bodies in *Chlamydomonas* may be prevented either physically, by the presence of a cell wall, or by the restricted occurrence of centrin within the fiber (see earlier). As with system II fibers, we anticipate several different functions for distal connecting fibers and for centrin itself in distal connecting fibers (see also Hoops et al., 1984, for a discussion of other functional properties of distal connecting fibers).

Finally, centrin has been found as a component of the flagellar transitional region in several taxa of green algae (Schulze et al., 1987; Melkonian et al., 1988; Sanders and Salisbury, 1989). This localization is most clearly seen using postembedding immuno-EM techniques (Figs. 6.7–6.12). With immunofluorescence it is very difficult to recognize the faint labeling at the flagellar bases (see Sanders and Salisbury, 1989), and in preembedding techniques the protein A–-labeled gold particle presumably cannot penetrate into the transitional region (McFadden et al., 1987; Figs. 6.14, 6.16, 6.17). It is not clear how widespread centrin is in the transitional region of flagellate green algae. So far 10 taxa were positive and none was negative. As with system II fibers and distal connecting fibers, the functional significance of centrin in the flagellar transitional region of green algae is not entirely clear. Sanders and Salisbury (1989) gave ultrastructural evidence that fibers of the stellate structure in *Chlamydomonas reinhardtii* contract upon flagellar shedding and displace the outer-doublet microtubules of the axoneme inward. They suggested that the resulting torsional and transverse shear forces sever axonemal microtubules immediately distal to the stellate structure. Since centrin is located in the stellate structure (the exact location is difficult to determine because of the small dimensions of the structure and the relatively large size of the gold particles), Sanders and Salisbury (1989) concluded that it is involved in microtubule severing during flagellar shedding. However, a flagellar autotomy mutant fa–1 (Lewin and Burrascano, 1983), which fails to shed its flagella, also contained centrin in the transitional region but was

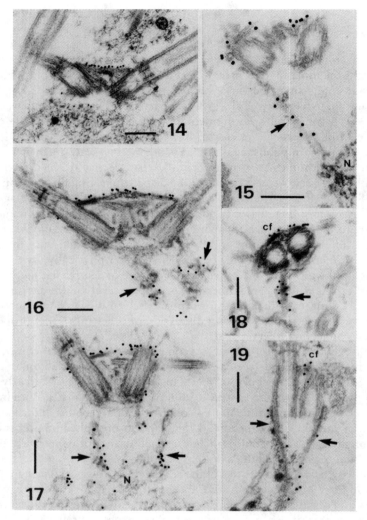

Figures 6.14 to 6.19. Pre-embedding immunoelectron microscopy of green algal cytoskeletons labeled with anti-centrin and protein A/gold. Bar = 0.25 μm.

(14) *Spermatozopsis similis*. The distal connecting fiber is heavily labeled.

(15) *Spermatozopsis exsultans*. Labeling is visible on portions of the distal connecting fibers and a system II fiber (arrow) that links the basal bodies to the nucleus (N).

(16, 17) *Hafniomonas reticulata*. Labeling is restricted to the distal connecting fiber and remnants of the system II fibers (arrows) that run to the nucleus (n).

(18) *Monostroma grevillei*, female gametes. The distal connecting fiber (cf) and a system II fiber (arrow) are antigenic.

(19) *Scourfieldia caeca*. Labeling is apparent on the distal connecting fiber (cf) and system II fibers (arrows) that arise from the proximal ends of both basal bodies.

Methods of fixation and immunolabeling are as described for *Spermatozopsis similis* in McFadden et al. (1987). Sources of other cultures: *Spermatozopsis exsultans*, Melkonian et al. (1987); *Hafniomonas reticulata*, Kreimer and Melkonian (1990); *Monostroma grevillei*, *Scourfieldia caeca*, Schulze et al. (1987).

unable to contract the fibers of the stellate structure to the same extent as wild-type cells (Sanders and Salisbury, 1989). Jarvik and Suhan (1991) have recently shown conspicuous ultrastructural defects in the transitional region of the vfl–2 mutant of *Chlamydomonas reinhardtii*. In place of the highly organized stellate structure of wild-type cells, the transitional region of the mutant contained only variable quantities of poorly organized electron-dense material. Nonetheless, vfl–2 cells are capable of effective flagellar shedding, indicating that shedding does not depend on properly assembled stellate structures. Since vfl–2 is known to be defective in centrin (see earlier), Jarvik and Suhan (1991) concluded that centrin is not involved in flagellar shedding of vfl–2 mutants. As an alternative function for centrin in the transitional region, Jarvik and Suhan (1991) proposed that stellate fiber contraction "reflects the fact that the fibers are normally in tension in order to hold the basal cylinders in place."

Occasionally, a faint and uniform anticentrin fluorescence was observed in flagella of *Chlamydomonas reinhardtii* (e.g., Huang et al., 1988a). Piperno et al. (1990) isolated two types of inner dynein arms from *Chlamydomonas* axonemes that, in addition to heavy chains, contained several polypeptides of lower molecular weight, including a 19-kD polypeptide that shows immunological cross-reactivity to centrin. Recently, Piperno and Ramanis (1991) showed that a specific type of inner dynein arm is apparently located exclusively in the proximal region of *Chlamydomonas* flagella. At present it is not clear how the low-molecular-weight polypeptides (including the putative centrin) distribute between proximal and distal types of inner dynein arms. The functional significance of the putative axonemal centrin remains to be evaluated.

Euglenophyceae

No published evidence exists on the occurrence of centrin in the Euglenophyceae. Our own unpublished immunofluorescence data indicate that a centrin homologue is present in *Euglena mutabilis*. Localization is restricted to two small dots associated with two basal bodies (Meinicke-Liebelt, unpublished observations).

Dinophyceae

There is a considerable body of knowledge about nonactin filaments in the Dinophyceae (Cachon and Cachon, 1981, 1985). Previous immunological evidence indicated that these filament bundles do not cross-react with actin, myosin, tubulin, or dynein. Some of the filament bundles are apparently involved in rapid motile responses. The piston of *Erythropsidinium pavillardi* (Warnowiidae) can be retracted and expanded within 200 ms (Greuet,

1976). The cell body of some dinoflagellates (especially the Leptodiscinae) is able to contract rapidly to some very unusual cell shapes (Cachon and Cachon, 1984a). Contractions are Ca^{2+} dependent but independent of external ATP, and a layer of 2–3-nm filaments is located beneath the cell surface that cannot be decorated with heavy meromyosin (Cachon and Cachon, 1984a). The ultrastructural changes observed in filaments upon cell body contraction suggest that a centrin homologue could be operative in these organisms. Similar types of filaments have been implicated in rapid movements associated with prey capture in *Kofoidinium* (Cachon and Cachon, 1985), in pusule contractility (Cachon et al., 1983), and in parasite–host interactions (Cachon and Cachon, 1984b). The longitudinal flagellum of some Dinophyceae is likewise capable of rapid retractions (see early light microscopy by Schütt, 1895; Metzner, 1929; Peters, 1929). The process of retraction in the longitudinal flagellum of *Ceratium tripos* has been studied in some detail by Maruyama (1981; 1982; 1985a,b). The 220-μm-long flagellum of this species retracts into the sulcus within 28 msec after mechanical stimulation (Maruyama, 1981). In reactivated models of the longitudinal flagellum, retraction required at least 10 μM Ca^{2+} but was independent of external ATP. A cross-striated fiber (R-fiber; Maruyama, 1982) that accompanies the axoneme over most of its length consists of fine filaments that supercoil upon flagellar retraction (Maruyama, 1982). A similar cross-striated fiber (striated strand or paraxonemal fiber) is present in most (all?) transverse flagella of dinoflagellates (for a review see Dodge, 1987).

Centrin was first localized in the Dinophyceae in the transverse flagella of three taxa by immunofluorescence and immunoblotting using a polyclonal antibody generated against centrin from the rhizoplast of *Tetraselmis striata* (Höhfeld et al., 1988; Fig. 6.27). Isolated transverse flagella of *Peridinium inconspicuum* were shown to undergo a rapid Ca^{2+}-induced contraction in the absence of external ATP. Höhfeld et al. (1988) identified a 21-kD protein as the immunoreactive antigen and concluded that centrin located in the paraxonemal fiber confers contractility to the transverse flagellum of dinoflagellates. Several possible functions for centrin in the transverse flagellum of dinoflagellates were suggested by Höhfeld et al. (1988): the centrin-containing paraxonemal fiber could be under tension and thus may be responsible for retaining the flagellum inside the cingulum. In addition, contraction of the paraxonemal fiber could cause inhibition of axonemal wave propagation, resulting in a stop response of the cell. Such stop responses occur frequently in dinoflagellates (Hand and Schmidt, 1975) and are analogous to the photophobic responses of green algae. In consequence, Höhfeld et al. (1988) designated centrin-mediated cell motility a "shock motility-system." Although centrin has not yet been demonstrated in the longitudinal flagellum of *Ceratium* spp., the characteristics of the contractile response in these flagella strongly suggest that the R-fiber of the longitudinal flagellum of *Ceratium tripos* also contains centrin.

Centrin is apparently widespread in the dinoflagellates. At least four other locations of centrin have been identified. Roberts and Roberts (1991) report that in several dinoflagellates a centrin fiber occurs near the cell apex in an area from which the subthecal microtubules radiate toward the cingulum. We have observed a similar fiber in *Woloszynskia pascheri* (Meinicke-Liebelt and Melkonian, unpublished observations). This fiber is unusual among the algal centrin-containing structures studied so far in that it has no apparent connection to the basal apparatus (see also centrin in Pedinellophyceae, later). Centrin also features prominently in several as yet unidentified structures of the basal apparatus of *Oxyrrhis marina,* in the striated flagellar collars and their connective in *Woloszynskia pascheri,* and in the proximal portion of the longitudinal flagellum of *Woloszynskia pascheri* (Meinicke-Liebelt, Höhfeld and Melkonian, unpublished observations). In all these cases, the functional significance of centrin has not been evaluated although, with respect to the flagellar collars, it appears possible that centrin may play a role in flagellar shedding analogous to the situation in green algae.

Chromophyta

In the heterokont algae, the first report of centrin as visualized by immunofluorescence came from Koutoulis et al. (1988) in their study of spine-scale reorientation in the peculiar pedinellophycean alga *Apedinella radians*. Swimming cells of *Apedinella radians* bear six elongate, cellulosic spine-scales that undergo a striking reorientation just prior to a change in the swimming direction of cells. Koutoulis et al. (1988) demonstrated the presence of three cytoskeletal systems that may be involved in this peculiar type of movement: microtubules, actin filaments, and centrin-containing filaments. The centrin-containing filaments form a six-pointed star when visualized with anticentrin immunofluorescence. The centrin-containing filaments link to six cytoplasmic plaques to which the spines are connected and to six cylindrical electron-dense caps that surround the microtubule triads of the tentacles at the plasma membrane (Koutoulis et al., 1988). This "centrin star" is apparently independent of the basal apparatus, and Koutoulis et al. (1988) propose that contraction of these centrin-containing fibers is (perhaps in conjunction with the actin cytoskeleton) responsible for the reorientation of spine scales. In addition to the "centrin star," a smaller, ringlike structure, possibly associated with the basal bodies, is immunoreactive to centrin (Koutoulis et al., 1988).

Since there is no other published evidence for the presence of centrin in chromophyte algae, we will present here some of our unpublished observa-

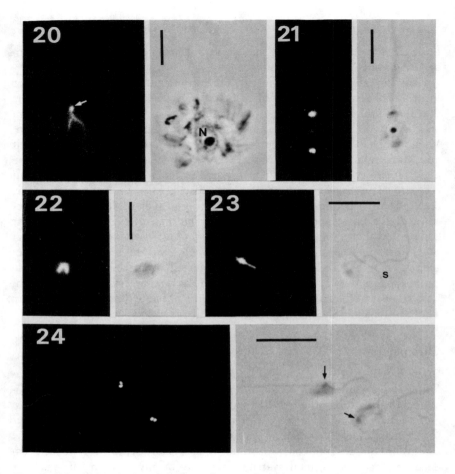

Figures 6.20 to 6.24. Anticentrin immunofluorescence and corresponding phase contrast images of permeablized chromophyte cells.

(20) *Synura petersenii*. A conecting fiber (arrow) joining the two basal bodies is antigenic, as is the rhizoplast/system II fiber, which shows two branches covering the anterior surface of the nucleus (N). Bar = 5 μm.

(21) *Pseudopedinella elastica*. Antigenicity is restricted to bright points at the anterior (flagellum is visible in phase contrast) and posterior of the cell. Some labeling is seen in the nucleolus; this is also visible in the preimmune control (not shown). Bar = 5 μm.

(22) *Pavlova lutheri*. Two distinct regions of antigenicity descending from the basal body region are apparent in all cells.

(23) *Ochromonas danica*. The bright, antigenic point near the basal bodies probably represents a contracted rhizoplast and/or basal body connecting fiber; a fine strand underlying the short flagellum(s) is also antigenic.

(24) *Fucus* sp., male gametes. Two antigenic spots are apparent in all cells; the corresponding regions (near the basal bodies) are indicated with arrows in the phase contrast image.

Methods of cell permeabilization and immunofluorescence are as described in Schulze et al. (1987). Culture sources: *Synura petersenii* SAG 120.79 (Schlösser, 1982); *Pseudopedinella elastica*, Heimann et al. (1989); *Pavlova lutheri*, SAG 926-1 (Schlösser, 1982); *Ochromonas danica*, SAG 933-7 (Schlösser, 1982); *Fucus* sp. collected from the coast of Helgoland (FRG) and gametes were released by flooding dried (overnight), fertile fronds with sterile seawater.

tions that show, based on immunofluorescence data, the widespread distribution of centrin or a centrin homologue in the Chromophyta (Figs. 6.20–6.24). Not surprisingly, an antigen recognized by anticentrin was found in the rhizoplasts (nucleus–basal body connectors) of *Synura petersenii* (Fig. 6.20). *Pavlova lutheri* (Fig. 6.22), and *Ochromonas danica* (in the contracted state; Fig. 6.23). Thus, we may conclude that rhizoplasts of the Chrysophyceae, Synurophyceae, and Prymnesiophyceae are presumably homologous structures to the system II fibers of green algae. Unfortunately, no immunoblot or immunoelectron microscope evidence exists as yet to support this conclusion further. In isolated cytoskeletons of *Ochromonas danica* we have found another presumptive centrin localization—namely, a fiber extending from the basal bodies parallel to and subtending the short axoneme (Fig. 6.23). This fiber may play a role in food capture. Food particle capture by flagella has been demonstrated for *Ochromonas danica* using image-enhanced video microscopy (Wetherbee and Andersen, 1991; see the review in Andersen, 1991). A "feeding pouch" forms around a large microtubular loop and provides a rapid means of engulfment of food particles (Andersen and Wetherbee, 1991). The microtubular loop forms by upward movement of microtubule(s) from the R3 microtubular root. The centrin fiber may be involved in some of these movements that are associated with the phagocytotic process in the Chrysophyceae. In this respect, we wish to emphasize that other methods for capturing food particles in chromophyte algae might also take advantage of centrin-mediated cell motility. In the Pedinellophyceae, food particles are captured by tentacles (Moestrup and Andersen, 1991). In *Pseudopedinella elastica* we have found two locations of centrin or an antigenic homologue of centrin on opposite ends of the cytoskeleton (Fig. 6.21): One end corresponds to the flagellar base (and the anterior tentacles); the other end, to the base of the posterior tentacles or the contractile stalk. Both the tentacles and the posterior contractile stalk can be retracted very rapidly. Koutoulis et al. (1988) also found centrin close to the cylindrical plaques that link the triad microtubules of the tentacles to the plasmamembrane. Finally, the haptonemata of some prymnesiophytes are apparently involved in food uptake (Kawachi and Inouye, personal communication). It is not known whether centrin is involved in the rapid Ca^{2+}-dependent coiling movements of such haptonemata.

We have localized centrin or a centrin homologue in vegetative cells and developing gametangia of *Ectocarpus siliculosus* (Katsaros et al., 1991) and in sperm cells of *Fucus* (Fig. 6.24). In cytoskeletons from *Fucus* spermatozoids we observed two small dots associated with the flagellar bases. A slightly different pattern was revealed in gametes of *Ectocarpus siliculosus*: two short rods connected to each other and to another, shorter fiber that extends toward the nucleus (Katsaros, Maier, and Melkonian, unpublished observations). It remains to be established with immunoelectron microscopy to which structures these patterns relate, but it is likely that rhizoplasts as well as basal body connectives (at least in *Ectocarpus* gametes) are involved.

Glaucocystophyta

Cyanophora paradoxa has been investigated for anticentrin labeling (Figs. 6.25, 6.26). A distinctive pattern of antigenicity was observed with four interconnected antigenic sites, two larger and two smaller globules, the larger of which consist of two to three subunits (Figs. 6.25, 6.26). Interestingly, the cyanelles remain associated with the basal apparatus when cells are treated with nonionic detergents, but it is not clear whether centrin plays any role in providing this connection. One of the smaller antigenic sites is close to the flagellar bases; the others appear to be located at some distance from them (Fig. 6.25). In summary, it appears that centrin or a centrin homologue is present in the Glaucocystophyta, but the labeling pattern in *Cyanophora paradoxa* is so unlike that found in any other algae (and is not readily comparable to known basal apparatus structures of *Cyanophora*) that functional considerations are precluded at present.

Cryptophyta

There are no published accounts of the occurrence of centrin in the Cryptophyta. We have studied anticentrin labeling both at the light and electron microscope level in two taxa of *Cryptomonas* and have devised experiments to analyze possible centrin functions (Schulze and Melkonian, unpublished observations). Labeling occurs in at least four different locations in the cells (Figs. 6.28–6.34). At the light microscope level, two strongly fluorescent broad bands run from the basal apparatus to the middle region of the cell, where they converge and terminate (Figs. 6.28, 6.29). A thin strand can be seen at this point that continues to the posterior end of the cell (Fig. 6.29). In addition, there is punctate labeling near the basal bodies, and a short, fluorescent strand extends transversely from the basal body region to the anterior dorsal periplast surface (Fig. 6.28, left micrograph). Immunoelectron microscopy has greatly facilitated identification of the respective antigenic sites (Figs. 6.30–6.34). The two strongly labeled broad bands correspond to the so-called rim fibers (Gillott and Gibbs, 1983). These represent thin sheets of filaments that underly the plasma membrane of the ventral furrow on the two sides bordering the ventral furrow opening (Figs. 6.30–6.32). Based on scanning electron microscope images, Kugrens et al. (1986) suggested that the ventral furrow of some *Cryptomonas* species may be capable of active movements, thereby closing or opening the groove. We have verified this and shown—using high-resolution light microscopy of in vivo immobilized cells (see Reize and Melkonian 1989 for description of the

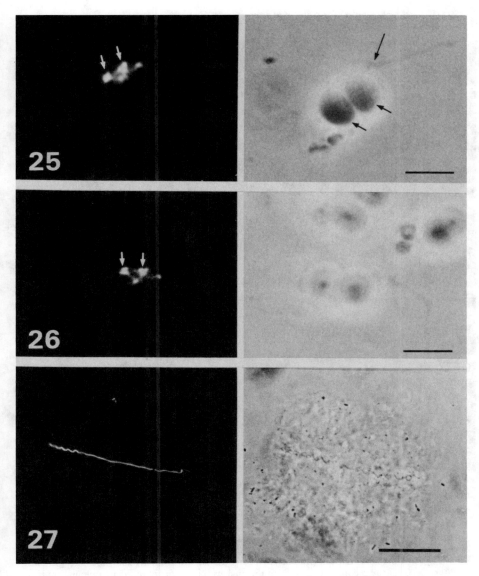

Figures 6.25 to 6.27. Anticentrin immunofluorescence and corresponding phase contrast images.

(25, 26) Isolated cytoskeletons of *Cyanophora paradoxa*. Four, interconnected antigenic sites are regularly apparent at the anterior end of all cells. Two sites (arrows, left micrographs) are larger and can be resolved as double structures (see Fig. 26). The two cyanelles (short arrows, phase contrast, Fig. 25) always remain associated with the cytoskeletons. Flagellar bases are indicated by the long arrow in the phase contrast image of Fig. 25. Bar = 5 μm. Methods are as described in Schulze et al. (1987); culture source is SAG 29.80 (Schlösser, 1982).

(27) *Gyrodinium resplendens*. The transverse flagellum is antigenic. Bar = 20 μm. Methods and culture as described in Höhfeld et al. (1988).

209

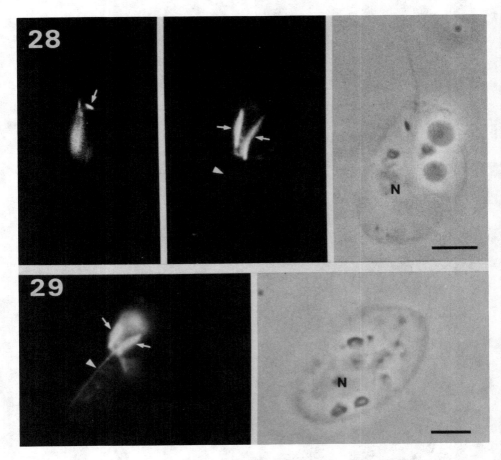

Figures 6.28 and 6.29. Anticentrin immunofluorescence and corresponding phase contrast images of permeabilized cells of *Cryptomonas* sp. Bar = 5 μm.

(28) Immunofluorescent image is shown in two focal planes. A bright region of antigenicity is visible in the region of the basal bodies (arrow, left micrograph); the rim fibers are also distinct (arrows, middle micrograph) and a faint reaction can be seen from the rhizostyle (arrowhead, middle micrograph) which runs alongside the nucleus (N).

(29) The rim fibers (arrows) and rhizostyle (arrowhead) are distinctly antigenic. Backgrouind fluorescence at the cell anterior and from the nucleus (N) is present in the preimmune control (not shown).

Methods of cell permeabilization and immunofluorescence are as described in Schulze et al. (1987). Culture source is SAG B 25.80 (Schlösser 1982).

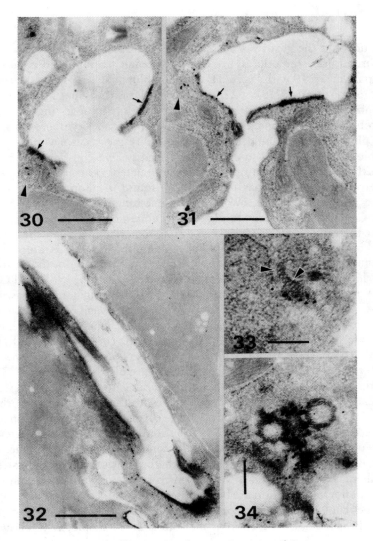

Figures 6.30 to 6.34. Postembedding immunoelectron microscopy of *Cryptomonas* sp. embedded in Lowicryl K4M resin and labeled with anticentrin and protein A/gold.

(30, 31) Transverse sections through cells with the furrow open (Fig. 30) and almost closed (Fig. 31). The rim fibers (arrows) and a fibrous structure associated with the microtubular arc of the rhizostyle (arrowhead) are labeled. Bar = 0.5 μm.

(32) Longitudinal section through a furrow showing heavy labeling of the rim fibers. Bar = 1.0 μm.

(33) High magnification of the rhizostyle showing labeling of the fibrous material associated with the microtubular arc (arrowheads). Bar = 0.25 μm.

(34) High magnification showing labeling of basal body-associated connecting fibers. Bar = 0.25 μm.

Methods of fixation and immunolabeling as described in Schulze et al. (1987). The strain of *Cryptomonas* sp. was isolated from Trenant, Cornwall, UK.

method) that the groove is normally open but can be rapidly closed upon perturbation of the cells (Reize and Melkonian, unpublished results; also compare Fig. 6.30 with Fig. 6.31). Immunoblot analyses of cytoskeletons of *Cryptomonas* spp. reveal a 22-kD protein that is recognized by anticentrin (Schulze and Melkonian, unpublished observations), suggesting that centrin is probably involved in closing the ventral groove of *Cryptomonas* species. The physiological significance of this motility mechanism is not clear: A phobic reaction; an aid in ejectisome discharge, which often accompanies perturbation of the cells; or a mechanism related to heterotrophic feeding (heterotrophic activity has recently been found in photosynthetic crypto-phytes; Kugrens and Lee, 1990) are possible functions of rim fiber contrac-tion. Centrin has also been localized with immunogold electron microscopy to the rhizostyle (Figs. 6.30, 6.31, 6.33), to material interconnecting the two basal bodies (Fig. 6.34) and to a fiber linking the basal apparatus with the vestibulum (not shown). Of some interest is the location in the rhizostyle. The rhizostyle of cryptophytes (for review see Moestrup, 1982, and Gillott, 1990) is a peculiar microtubular flagellar root that originates alongside one of the basal bodies and extends toward the posterior end of the cell. In many cryptophytes each of the rhizostyle microtubules bears a winglike lamellar projection (Mignot et al., 1968; Roberts et al., 1981; Gillot and Gibbs, 1983). These wings are absent in some members of the genus *Cryptomonas* (Roberts, 1984). In our material the wings are also absent (Fig. 6.33), but centrin is present in a fiber that runs parallel to the rhizostyle, closely associated with the rhizostyle microtubules, to the posterior end of the cell (Fig. 6.29). In cross sections through the rhizostyle, this fiber is apparently located either toward the outside of the convex band of microtubules (Fig. 6.33) or, more often, to the inside of the convex array of microtubules (Figs. 6.30, 6.31). It needs to be established whether this variability indicates that the fiber is twisted around the row of microtubules or if different functional states of fiber contraction are being observed. (If linked to the microtubules, the fiber could cause the row of microtubules to change shape upon contrac-tion!) In general, the function of the rhizostyle is completely unknown.

Rhodophyta

Although flagellate cells do not occur in the Rhodophyta, recent evidence from sequence analysis of r-DNA indicates that the group is presumably not primitively aflagellate (Bhattacharya et al., 1990; see also discussion in Melkonian, 1991). In support of this conclusion there is preliminary evidence for the presence of centrin or a centrin homologue in the group (Dassler et

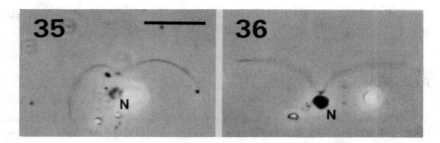

Figures 6.35 and 6.36. Detergent-extracted cytoskeleton–nuclear complexes of *Dunaliella bioculata*. Extended (Fig. 35; ≤1.5 × 10⁻⁸ *M* Ca²⁺) and contracted (Fig. 36; ≥10⁻⁷ *M* Ca²⁺) nucleus–basal body connectors. N = nucleus. Bar = 10 μm. Methods for the isolation of viable cytoskeletons are as described in McFadden et al. (1987). Culture source is SAG 19-4 (Schllösser 1982).

al., 1990). It will be interesting to learn where centrin is located in the red algal cell.

Conclusion

A novel type of cell motility mechanism has been discovered in the algae. It is characterized by rapid contractions of an intracellular filamentous structure within less than 20 ms and by much slower reextension of the filamentous structure to its original length (in the range of seconds up to 1 hour). Contraction is Ca²⁺ induced, but independent of external ATP, reextension requires removal of Ca²⁺ from the structure and is energy dependent. The major protein of the contractile filaments, centrin, is a Ca²⁺-modulated 20-kD phosphoprotein of the EF-hand protein family that presumably undergoes large conformational changes upon Ca²⁺ binding. It has been suggested that the conformational changes of centrin lead to filament contraction by twisting and supercoiling of the filaments. Centrin-mediated cell motility presumably occurs in all algal groups; it is most prominent in flagellate cells, where centrin is almost always a component of the basal apparatus—the equivalent of the eukaryotic centrosome in flagellate cells. The functional significance of centrin-mediated cell motility—even in the comparatively well-studied green algal systems—remains elusive. A general proposal, however, is that centrin-mediated cell motility may represent an ancient "shock motility system" (see Höhfeld et al., 1988) that enables cells to react quickly to environmental stress. More insight into the molecular and physiological function of centrin-mediated cell motility may be expected once centrin-containing filaments reassembled from the monomer protein can be reactivated for contraction cycles in vitro.

Acknowledgments

We thank the members of our laboratory for helpful discussions and some help with experimental data. This study was supported by grants from the Deutsche Forschungsgemeinschaft (Me 658/3–4, Me 658/3–5). PLB thanks the Alexander von Humboldt Foundation for a supporting fellowship, and CHRK thanks the Stiftung Volkswagenwerk and the Ministry of Research and Technology of Greece for support of his visit to Cologne.

References

Amos, W. B. 1971. Reversible mechanochemical cycle in the contraction of *Vorticella*. *Nature* 229:127–128.
Amos, W. B. 1972. Structure and coiling of the stalk in the peritrich ciliates *Vorticella* and *Carchesium*. *J. Cell Sci.* 10:95–122.
Amos, W. B. 1975. Contraction and calcium binding in vorticellid ciliates. In R. E. Stephens and S. Inoue (eds.), *Molecules and Cell Movement,* Raven Press, New York, pp. 411–436.
Amos, W. B., Routledge, L. M., and Yew, F. F. 1975. Calcium-binding proteins in a vorticellid contractile organelle. *J. Cell Sci.* 19:203–213.
Andersen, R. A. 1991. The cytoskeleton of chromophyte algae. *Protoplasma* (in press).
Andersen, R. A., and Wetherbee, R. 1991. Microtubules of the flagellar apparatus are active during prey capture in the chrysophycean alga *Epipyxis pulchra*. *Cell Motil. Cytoskel.* (submitted).
Bannister, L. H., and Tatchell, E. C. 1968. Contractility and the fiber systems of *Stentor coeruleus*. *J. Cell Sci.* 3:295–308.
Baron, A. T., Greenwood, T. M., and Salisbury, J. L. 1991. Localization of the centrin-related 165,000-M_r protein of ptK$_2$ cells during the cell cycle. *Cell Motil. Cytoskel.* 18:1–14.
Baron, A. T., and Salisbury, J. L. 1987. Identification and localization of a novel cytoskeletal centrosome associated protein, "centrin," in PtK2 cells. *J. Cell Biol.* 105:205a.
Baron, A. T., and Salisbury, J. L. 1988. Identification and localization of a novel cytoskeletal, centrosome-associated protein in PtK2 cells. *J. Cell Biol.* 107:2669–2678.
Bhattacharya, D., Elwood, H. J., Goff, L. J., and Sogin, M. L. 1990. Phylogeny of *Graciliaria lemaneiformis* (Rhodophyta) based on sequence analysis of its small subunit ribosomal RNA coding region. *J. Phycol.* 26:181–186.
Cachon, J., and Cachon, M. 1981. Movement by non-actin filament mechanisms. *BioSystems* 14:313–326.

Cachon, J., and Cachon, M. 1984a. An unusual mechanism of cell contraction: Leptodiscinae dinoflagellates. *Cell Motil.* 4:41–45.

Cachon, J., and Cachon M. 1984b. A new Ca^{2+}-dependent function of flagellar rootlets in Dinoflagellates: The releasing of a parasite from its host. *Biol. Cell* 52:61–76.

Cachon, J., and Cachon, M. 1985. Non-actin filaments and cell contraction in *Kofoidinium* and other dinoflagellates. *Cell Motil.* 5:1–15.

Cachon, J., Cachon, M., and Boillot, A. 1983. Flagellar rootlets as myonemal elements for pusule contractility in dinoflagellates. *Cell Motil.* 3:61.

Carasso, N., and Favard, P. 1966. Mis en evidence du calcium dans les myonèmes pédonculaires de ciliés péritriches. *J. Microsc.* 5:759–770.

Coling, D. E., and Salisbury, J. L. 1987. Purification and characterization of centrin, a novel calicum-modulated contractile protein. *J. Cell Biol.* 105:205a.

Dangeard P. A. 1901. Etude comparative de la zoospore et du spermatozoïde. *C.R. Hebd. Seances Acad. Sci.* 132:859–861.

Dassler, C. L., Scott, J., and Salisbury, J. L. 1990. Centrin in "primitive" eucaryotes: Centrin homologues of the Rhodophyta. *J. Cell Biol.* 111:182a.

Dodge, J. D. 1987. Dinoflagellate ultrastructure and complex organelles: A. General ultrastructure. In F. J. R. Taylor (ed.), *The Biology of Dinoflagellates.* Blackwell, Oxford, pp. 93–119.

Doflein, F. 1918. Beiträge zur Kenntnis von Bau und Teilung der Protozoenkerne. *Zool Anz.* 46:289–306.

Engelmann, T. W. 1875. Contractilität und Doppelbrechung. *Pflügers Arch. Eur. J. Physiol.* 11:432–464.

Entz, G. 1892. Die elastischen und contractilen Elemente der Vorticellen. *Math. Naturw. Ber. Ung.* 10:1–48.

Ettienne, E. M. 1970. Control of contractility in *Spirostomum* by dissociated calcium ions. *J. Gen. Physiol.* 56:168–179.

Ettienne, E. M., and Seltsky, M. 1974. The antagonistic agents on contraction and re-extension in the ciliate *Spirostomum ambiguum. J. Cell Sci.* 16:377–383.

Ettl, H., and Moestrup, Ø. 1980. Light and electron microscopical studies on *Hafniomonas* gen. nov. (Chlorophyceae, Volvocales), a genus resembling *Pyramimonas* (Prasinophyceae). *Pl. Syst. Evol.* 135:177–210.

Gillott, M. 1990. Phylum Cryptophyta. In L. Margulis, J. O. Corliss, M. Melkonian, and D. J. Chapman (eds.), *Handbook of Protoctista,* Jones and Bartlett, Boston, pp. 139–151.

Gillott, M., and Gibbs, S. P. 1983. Comparison of the flagellar rootlets and periplast in two marine cryptomonads. *Can. J. Bot.* 61:1964–1978.

Goodenough, U. W. 1983. Motile detergent-extracted cells of *Tetrahymena* and *Chlamydomonas. J. Cell Biol.* 96:1610–1621.

Greuet, C. 1976. Organisation et fonctionnement du tentacule postérieur d'*Erythropsidinium*, Pérdinien Warnowiidae, *J. Microsc. Biol. Cell.* 27:13a.

Hamburger, C. 1905. Zur Kenntnis der *Dunaliella salina* und einer Amöbe aus Salinenwasser von Cagliari. *Arch. Protistenkd.* 6:111–130.

Hand, W. G., and Schmidt, J. A. 1975. Phototactic orientation by the marine dinoflagellate *Gyrodinium dorsum* Kofoid. II. Flagellar activity and overall response mechanisms. *J. Protozool.* 22:494–498.

Harris, E. H. 1989. *The Chlamydomonas Sourcebook: A Comprehensive Guide to Biology and Laboratory Use.* Academic Press, San Diego, 780 pp.

Hiraoka, L., Golden, W., and Magnuson, T. 1989. Spindle pole organization during early mouse development. *Dev. Biol.* 133:24–36.

216   Melkonian, Beech, Katsaros, and Schulze

Hoffman, L. R. 1973. Fertilization in Oedogonium. I. Plasmogamy. *J. Phycol.* 9:62–84.

Hoffmann-Berling, H. 1958. Der Mechanismus eines neuen, von der Muskelkontraktion verschiedenen Kontraktionszyklus. *Biochim. Biophys. Acta* 27:247–255.

Höhfeld, I., Otten, J., and Melkonian, M. 1988. Contractile eukaryotic flagella: Centrin is involved. *Protoplasma* 147:16–24.

Hoops, H. J., Wright, R. L., Jarvik, J. W., and Witman, G. B. 1984. Flagellar waveform and rotational orientation in a *Chlamydomonas* mutant lacking normal striated fibers. *J. Cell Biol.* 98:818–824.

Huang, B., and Mazia, D. 1975. Microtubules and filaments in ciliate contractility. In R. E. Stephens and S. Inouye (eds.), *Molecules and Cell Movement,* Raven Press, New York, pp 389–410.

Huang, B., Mengersen, A., and Lee, V. D. 1988b. Molecular cloning of cDNA for caltractin, a basal body-associated Ca^{2+}-binding protein: Homology in its protein sequence with calmodulin and the yeast CDC31 gene product. *J. Cell Biol.* 107:133–140.

Huang, B., and Pitelka, D. R. 1973. The contractile process in the ciliate *Stentor coeruleus. J. Cell Biol.* 57:704–723.

Huang, B., Watterson, D. W., Lee, V. D., and Schibler, M. J. 1988a. Purification and characterization of a basal body-associated Ca^{2+}-binding protein. *J. Cell Biol.* 107:121–131.

Hyams, J. S., and Borisy, G. G. 1978. Isolated flagellar apparatus of *Chlamydomonas:* Characterization of forward swimming and alteration of waveform and reversal of motion by calcium ions in vitro. *J. Cell Sci.* 33:235–253.

Ishida, H., and Shigenaka, Y. 1988. Cell model contraction in the ciliate *Spirostomum. Cell Motil. Cytoskel.* 9:278–282.

Jarvik, J. W., and Suhan, J. P. 1991. The role of the flagellar transition region: Inferences from the analysis of the *Chlamydomonas* mutant vfl–2. *J. Cell Sci.* (in press).

Jones, A. R., Jahn, T. L., and Fonseca, J. R. 1970. Contraction of protoplasm. IV. Cinematographic analysis of the contraction of some peritrichs. *J. Cell. Physiol.* 75:9–20.

Kater, J. M. 1929. Morphology and division of *Chlamydomonas* with reference to the phylogeny of the flagellar neuromotor system. *Univ. Calif. Publ. Zool.* 133:125–168.

Katsaros, C., Kreimer, G., and Melkonian, M. 1991. Localization of tubulin and a centrin-homologue in vegetative cells and developing gametangia of *Ectocarpus siliculosus* (Dillw.) Lyngb. (Phaeophyceae, Ectocarpales). *Bot. Acta* 104:87–92.

Koutoulis, A., McFadden, G. I., and Wetherbee, R. 1988. Spinescale reorientation in *Apedinella radians* (Pedinellales, Chrysophyceae): The microarchitecture and immunocytochemistry of the associated cytoskeleton. *Protoplasma* 147:25–41.

Kreimer, G., and Melkonian, M. 1990. Reflection confocal laser scanning microscopy of eyespots in flagellated green algae. *Eur. J. Cell Biol.* 53:101–111.

Kristensen, B. T., Engdahl Nielsen, L., and Rostgaard, J. 1974. Variations in myoneme birefringence in relation to length changes in *Stentor coeruleus. Exp. Cell Res.* 85:127–135.

Kugrens, P., and Lee, R. E. 1990. Ultrastructural evidence for bacterial incorporation and myxotrophy in the photosynthetic cryptomonad *Chroomonas pochmanni* Huber-Pestalozzi (Cryptomonadida). *J. Protozool.* 37:263–267.

Kugrens, P., Lee, R. E., and Andersen, R. A. 1986. Cell form and surface patterns in *Chroomonas* and *Cryptomonas* cells (Cryptophyta) as revealed by scanning electron microscopy. *J. Phycol.* 22:512–522.

Lechtreck, K.-F., McFadden, G. I., Melkonian, M. 1989. The cytoskeleton of the naked green flagellate *Spermatozopsis similis:* Isolation, whole mount electron microscopy, and preliminary biochemical and immunological characterization. *Cell Motil. Cytoskel.* 14:552–561.

Lechtreck, K.-F., and Melkonian, M. 1991. An update on fibrous flagellar roots in green algae. *Protoplasma* (in press).

LeDizet, M., and Piperno, G. 1986. Cytoplasmic microtubules containing acetylated α-tubulin in *Chlamydomonas reinhardtii:* Spatial arrangement and properties. *J. Cell Biol.* 103:13–22.

Lewin, R. A., and Burrascano, C. 1983. Another new kind of *Chlamydomonas* mutant, with impaired flagellar autotomy. *Experientia* 39:1397–1398.

Lewin, R. A., and Lee, K. W. 1985. Autotomy of algal flagella: Electron microscope studies of *Chlamydomonas* (Chlorophyceae) and *Tetraselmis* (Prasinophyceae). *Phycologia* 24:311–316.

Manton, I. 1952. The fine structure of plant cilia. *Symp. Soc. Exp. Biol.* 6:306–319.

Manton, I. 1956. Plant cilia and associated organelles. In D. Rudnick (ed.), *Cellular Mechanisms in Differentiation and Growth.* Princeton University Press, Princeton, pp. 61–71.

Manton, I. 1965. Some phyletic implications of flagellar structure in plants. In R. D. Preston (ed.), *Advances in Botanical Research,* Vol. 2. Academic Press, London, pp. 1–34.

Manton, I., Parke, M. 1965. Observations on the fine structure of two species of *Platymonas* with special reference to flagellar scales and the mode of origin of the theca. *J. Mar. Biol. Assoc. U.K.* 45:743–754.

Martindale, V. E., and Salisbury, J. L. 1990. Phosphorylation of algal centrin is rapidly responsive to changes in the external milieu. *J. Cell Sci.* 96:395–402.

Maruyama, T. 1981. Motion of the longitudinal flagellum in *Ceratium tripos* (Dinoflagellida)—a retractile flagellar motion. *J. Protozool.* 28:328–336.

Maruyama, T. 1982. Fine structure of the longitudinal flagellum in *Ceratium tripos,* a marine dinoflagellate. *J. Cell Sci.* 58:109–123.

Maruyama, T. 1985a. Extraction model of the longitudinal flagellum of *Ceratium tripos* (Dinoflagellida): Reactivation of flagellar retraction. *J. Cell Sci.* 75:313–328.

Maruyama, T. 1985b. Ionic control of the longitudinal flagellum in *Ceratium tripos* (Dinoflagellida). *J. Protozool.* 32:196–210.

Mattox, K. R., and Stewart, K. D. 1984. In D. E. G. Irvine and D. M. John (eds.), *Systematics of the Green Algae.* Academic Press, London, pp. 29–72.

McFadden, G. I., and Melkonian, M. 1986. Golgi apparatus activity and membrane flow during scale biogenesis in the green flagellate *Scherffelia dubia* (Prasinophyceae). I. Flagellar regeneration. *Protoplasma* 130:186–198.

McFadden, G. I., Schulze, D., Surek, B., Salisbury, J. L., and Melkonian, M. 1987. Basal body reorientation mediated by a Ca^{2+}-modulated contractile protein. *J. Cell Biol.* 105:903–912.

Melkonian, M. 1979. An ultrastructural study of the flagellate *Tetraselmis cordiformis* Stein (Chlorophyceae) with emphasis on the flagellar apparatus. *Protoplasma* 98:139–151.

Melkonian, M. 1980a. Ultrastructural aspects of basal body associated fibrous structures in green algae: A critical review. *BioSystems* 12:85–104.

Melkonian, M. 1980b. Flagellar roots, mating structure and gametic fusion in the green alga *Ulva lactuca* (Ulvales). *J. Cell Sci.* 46:149–169.

Melkonian, M. 1984. Flagellar apparatus ultrastructure in relation to green algal classification. In D. E. G. Irvine and D. M. John (eds.), *Systematics of the Green Algae.* Academic Press, London, pp. 73–120.

Melkonian, M. 1989. Centrin-mediated motility: A novel cell motility mechanism in eukaryotic cells. *Bot. Acta* 102:3–4.

Melkonian, M. 1990. Prasinophyceae. In L. Margulis, J. O. Corliss, M. Melkonian, and D. J. Chapman (eds.), *Handbook of Protoctista.* Bartlett and Jones, Boston, pp. 600–607.

Melkonian, M. 1991. Systematics and evolution of the algae. *Prog. Bot.* 52:271–307.

Melkonian, M., Becker, B., and Becker, D. 1991. Scale formation in algae. *J. Electron Microsc. Tech.* 17:165–178.

Melkonian, M., McFadden, G. I., Reize, I. B., and Preisig, H. R. 1987. A light and electron microscopic study of the quadriflagellate green alga *Spermatozopsis exsultans. Pl. Syst. Evol.* 158:47–61.

Melkonian, M., and Preisig, H. R. 1984. Ultrastructure of the flagellar apparatus in the green flagellate *Spermatozopsis similis. Pl. Syst. Evol.* 146:145–162.

Melkonian, M., and Preisig, H. R. 1986. A light and electron microscopic study of *Scherffelia dubia,* a new member of the scaly green flagellates (Prasinophyceae). *Nord. J. Bot.* 6:235–256.

Melkonian, M., Schulze, D., McFadden, G. I., and Robenek, H., 1988. A polyclonal antibody (anti-centrin) distinguishes between two types of fibrous flagellar roots in green algae. *Protoplasma* 144:56–61.

Metzner, P. 1929. Bewegungsstudien an Peridineen. *Z. Bot.* 22:225–265.

Mignot, J.-P., Joyon, L., and Pringsheim, E. G. 1968. Complements a l'etude cytologique des cryptomonadines. *Protistologica* 4:493–506.

Moestrup., Ø. 1978. On the phylogenetic validity of the flagellar apparatus in green algae and other chlorophyll a and b containing plants. *BioSystems* 10:117–144.

Moestrup, Ø. 1982. Flagellar structure in algae: A review, with new observations particularly on the Chrysophyceae, Phaeophyceae (Fucophyceae), Euglenophyceae, and *Reckertia. Phycologia* 21:427–528.

Moestrup, Ø. and Andersen, R. A. 1991. Organization of heterotrophic heterokonts. In D. J. Patterson and J. Larsen (eds.), *The Biology of Free-Living Heterotrophic Flagellates,* Clarendon Press, Oxford, in press.

Moncrief, N. D., Kretsinger, R. H. and Goodman, M. 1990. Evolution of EF-hand calcium-modulated proteins. I. Relationships based on amino acid sequences. *J. Mol. Evol.* 30:522–562.

O'Kelly, C. J., and Floyd, G. L. 1984. Flagellar apparatus absolute orientations and the phylogeny of the green algae. *BioSystems* 16:227–252.

Parke, M., and Manton, I. 1965. Preliminary observations on the fine structure of *Prasinocladus marinus. J. Mar. Biol. Assoc. U.K.* 45:525–536.

Peters, H. 1929. Über Orts- und Geisselbewegung bei marinen Dinoflagellaten. *Arch. Protistenkd.* 67:291–321.

Pickett-Heaps, J. D. 1971. Reproduction by zoospores in *Oedogonium.* I. Zoosporogenesis. *Protoplasma* 72:275–314.

Pickett-Heaps, J. D. 1975. *Green Algae: Structure, Reproduction and Evolution in Selected Genera.* Sinauer, Sunderland, MA, 605 pp.

Piperno, G., and Ramanis, Z. 1991. The proximal portion of *Chlamydomonas* flagella contains a distinct set of inner dynein arms. *J. Cell Biol.* 112:701–709.

Piperno, G., Ramanis, Z., Smith, E. F. and Sale, W. S. 1990. Three distinct inner dynein arms in *Chlamydomonas* flagella: Molecular composition and location in the axoneme. *J. Cell Biol.* 110:379–389.

Pitelka, D. R. 1969. Fibrillar systems in protozoa. *Res. Protozool.* 3:279–388.

Preisig, H. R., and Melkonian, M. 1984. A light and electron microscopical study of the green flagellate *Spermatozopsis similis* sp. nov. *Pl. Syst. Evol.* 146:57–74.

Randall, J. T., and Jackson, S. F. 1958. Fine structure and function in *Stentor polymorphus*. *J. Biophys. Biochem. Cytol.* 4:807–830.

Reize, I. B., and Melkonian, M. 1987. Flagellar regeneration in the scaly green flagellate *Tetraselmis striata* (Prasinophyceae): Regeneration kinetics and effect of inhibitors. *Helgol. wiss. Meeresunters.* 41:149–164.

Reize, I. B., and Melkonian, M. 1989. A new way to investigate living flagellated/ ciliated cells in the light microscope: Immobilization of cells in agarose. *Bot. Acta* 102:145–151.

Ricketts, T. R. 1979. The induction of synchronous cell division in *Platymonas striata* Butcher (Prasinophyceae). *Br. phycol. J.* 14:219–223.

Ringo, D. L. 1967. Flagellar motion and fine structure of the flagellar apparatus in *Chlamydomonas reinhardtii*. *J. Cell Biol.* 33:543–571.

Robenek, H., and Melkonian, M. 1979. Rhizoplast–membrane associations in the flagellate *Tetraselmis cordiformis* Stein (Chlorophyceae) revealed by freeze-fracture and thin sections. *Arch. Protistenkd.* 122:340–351.

Roberts, K. R. 1984. Structure and significance of the cryptomonad flagellar apparatus. I. *Cryptomonas ovata* (Cryptophyta). *J. Phycol.* 20:590–599.

Roberts, K. R., and Roberts, J. E. 1991. The flagellar apparatus and cytoskeleton of the Dinoflagellates: a comparative overview. *Protoplasma* (in press).

Roberts, K. R., Stewart, K. D., and Mattox, K. R. 1981. The flagellar apparatus of *Chilomonas paramecium* and its comparison with certain zooflagellates. *J. Phycol.* 17:159–167.

Routledge, L. M. 1978. Calcium-binding proteins in the vorticellid spasmoneme. *J. Cell Biol.* 77:358–370.

Routledge, L. M., Amos, W. B., Gupta, B. L., Hall, T. A. and Weis-Fogh, T. 1975. Microprobe measurements of calcium-binding in the contractile spasmoneme of a vorticellid. *J. Cell Sci.* 19:195–201.

Routledge, L. M., Amos, W. B., Yew, F. F. and Weis-Fogh, T. 1976. New calcium-binding contractile proteins. In R. Goldman, T. Pollard, and J. Rosenbaum (eds.), *Cell Motility,* Cold Spring Harbor Laboratory, pp. 93–113.

Rüffer, U., and Nultsch, W. 1985. High-speed cinematographic analysis of the movement of *Chlamydomonas*. *Cell Motil.* 5:251–263.

Sabatini, D. D., Bensch, K., and Barnett, R. J. 1963. Cytochemistry and electron microscopy. *J. Cell Biol.* 17:19–58.

Salisbury, J. L. 1982. Calcium-sequestering vesicles and contractile flagellar roots. *J. Cell Sci.* 58:433–443.

Salisbury, J. L. 1983. Contractile flagellar roots: The role of calcium. *J. Submicrosc. Cytol.* 15:105–110.

Salisbury, J. L. 1989a. Algal centrin-calcium-sensitive contractile organelles. In A. W. Coleman, L. J. Goff, and J. R. Stein-Taylor (eds.), *Algae as Experimental Systems,* Alan R. Liss, New York, pp. 19–37.

Salisbury, J. L. 1989b. Centrin and the algal flagellar apparatus. *J. Phycol.* 25:201–206.

Salisbury, J. L., Baron, A., Surek, B., and Melkonian, M. 1984. Striated flagellar roots: Isolation and partial characterization on a calcium-modulated contractile organelle. *J. Cell Biol.* 99:962–970.

Salisbury, J. L., Baron, A. T., Coling, D. E., Martindale, V. E., and Sanders, M. A. 1986. Calcium-modulated contractile proteins associated with the eucaryotic centrosome. *Cell Motil.* 6:193–197.

Salisbury, J. L., Baron, A. T., and Sanders, M. A. 1988. The centrin-based cytoskeleton of *Chlamydomonas reinhardtii*: Distribution in interphase and mitotic cells. *J. Cell Biol.* 107:635–641.

Salisbury, J. L., and Floyd, G. L. 1978. Calcium-induced contraction of the rhizoplast of a quadriflagellate green alga. *Science* 202:975–977.

Salisbury, J. L., and Greenwood, T. M. 1990. Molecular conservation of the eucaryotic centrosome: Assembly of algal centrin into functional vertebrate centrosomes. *J. Cell Biol.* 111:181a.

Salisbury, J. L., Sanders, M. A., and Harpst, L. 1987. Flagellar root contraction and nuclear movement during flagellar regeneration in *Chlamydomonas reinhardtii*. *J. Cell Biol.* 105:1799–1805.

Salisbury, J. L., Swanson, J. A., Floyd, G. L., Hall, R., and Maihle, N. J. 1981. Ultrastructure of the flagellar apparatus of the green alga *Tetraselmis subcordiformis* with special consideration given to the function of the rhizoplast and rhizanchora. *Protoplasma* 107:1–11.

Sanders, M. A., and J. L. Salisbury. 1989. Centrin-mediated microtubule severing during flagellar excision in *Chlamydomonas reinhardtii*. *J. Cell Biol.* 108:1751–1760.

Schlösser, U. G. 1982. Sammlung von Algenkulturen. *Ber. Deutsch. Bot. Ges.* 95:181–276.

Schnepf, E., and Maiwald, M. 1970. Halbdesmosomen bei Phytoflagellaten. *Experientia* 26:1343.

Schulze, D., Robenek, H., McFadden, G. I., and Melkonian, M. 1987. Immunolocalization of a Ca^{2+}-modulated contractile protein in the flagellar apparatus of green algae: the nucleus–basal body connector. *Eur. J. Cell Biol.* 45:51–61.

Schütt, F. 1895. Die Peridineen der Plankton-Expedition. I. Studien über die Zellen der Peridineen. *Ergebn. Plankton-Exped.* (*Kiel und Leipzig*) 4:1–170.

Stewart, K. D., and Mattox, K. R. 1975. Comparative cytology, evolution and classification of the green algae with some consideration of the origin of other organisms with chlorophyll a and b. *Bot. Rev.* 41:104–135.

van Leeuwenhoek, A. 1676. Letter. *Phil. Trans. Roy. Soc. Lond. B* 12:133.

Weis-Fogh, T., and Amos, W. B. 1972. Evidence for a new mechanism of cell motility. *Nature* 236:301–304.

Wetherbee, R., and Andersen, R. A. 1991. Flagella of chrysophycean algae play an active role in prey capture and selection: direct observations on *Epipyxis pulchra* and *Ochromonas danica* using image enhanced video microscopy. *Cell Motil. Cytoskel.* (submitted).

White, R. B., and Brown, D. L. 1981. ATPase activities associated with the flagellar basal apparatus of *Polytomella*. *J. Ultrastruct. Res.* 75:151–161.

Wright, R. L., Salisbury, J., and Jarvik, J. W. 1985. A nucleus–basal body connector in *Chlamydomonas reinhardtii* that may function in basal body localization or segregation. *J. Cell Biol.* 101:1903–1912.

Wright, R. L., Adler, S. A., Spanier, J. G., and Jarvik, J. W. 1989. Nucleus–basal body connector in *Chlamydomonas*: Evidence for a role in basal body segregation and against essential roles in mitosis or in determining cell polarity. *Cell Motil. Cytoskel.* 14:516–526.

Yogosawa-Ohara, R., Sizaki, T., and Shigenaka, Y. 1985. Twisting contraction mechanism of a heterotrichous ciliate, *Spirostomum ambiguum.* 2. Role of longitudinal microtubular sheet. *Cytobios* 44:215–230.

Zimmermann, W. 1925. Helgoländer Meeresalgen I–VI. *Wiss. Meeresunters. Abt. Helgoland* 16:1–25.

Taxonomic Index

Author Index

Subject Index